My Father– The Mapmaker, His Life and His art

"The Odyssey of the 95th Victory Division World War II"

Written by Lynda Clark Porter

Art by Richard E. Clark
with illustrations by Lynda Clark Porter

Lynda Clark Porter
LPorterArt@aol.com

Library of Congress Cataloging in Publication Data

Porter, Lynda Cark, Author

My Father – The Mapmaker, His Life and His Art
"The Odyssey of the 95th Victory Division World War II"

Summary: This book tells the true story of a World War II Veteran and Patriot, Richard E. Clark. His life journey is told through the tapestry of his vibrant artworks, his heartfelt speeches, his vivid descriptions, and through the voice of his loving daughter. While a soldier in World War II, Richard documented his personal journey in the 95th Victory Division with his maps, sketches and watercolors often painted under nighttime skies, lit by the dim light and din of distant bombings.

[1. Odyssey of the 95th Victory Division, World War II. 2. Cartography of World War II. 3. World War II Art and Paintings. 4. Iron Men of Metz. 5. Richard E. Clark.]

ISBN 978-0-9978111-1-7 (paperback)
Typesetting, Composition, Digital set-up and design: Rosanna I. Porter, Raisykinder Publishing.

On the cover: The Illustration depicts the dramatic story of Richard E. Clark painting a bombing scene in the distance, during World War II. The scene tells the story of his war time journey through his thoughts, experiences and artistic talents. See page 69. Lynda Clark Porter painted the illustration of her father sitting and painting during the war.

This book was written in honor of my parents,
Richard Edward and Jessie Marie Clark.

This book is dedicated to my husband Daniel,
and our sons, Nathaniel and Benjamin;
my sister, Rebecca and her husband, Jerry
and the descendants of Richard E. Clark.

Special Acknowledgement and gratitude to Jerry L. Latta for photography, cataloguing of photos and photography editing; Rebecca L. Latta for support, help with sorting through volumes of family art, photos, memories and writings; and Rosanna I. Porter for support, advice, digital design, layout and publication preparation. Daniel J. Porter for his unending patience, much needed assistance and support throughout the process of writing and assembling this book written for the preservation of good deeds done and memories kept. All of these are voiced with much love and appreciation for the generosity of family.

Richard E. Clark, "USA Troops Arriving in Metz"

TABLE OF CONTENTS

Richard E. Clark (pictured in back row second from the right) working in the field with the cartography team.

Introduction

Through a Daughter's Eyes

Richard Edward Clark, World War II

My father was a World War II Veteran. As I became more familiar with the man who was my father, and the aspects about him that for most of my life I had not understood, or thought to understand; I came to realize that his story needed to be shared. Unlike many stories of war veterans and the impact that World War II had on their lives, my father had an intrinsic talent that made his viewpoint unique. He was an artist. He had learned in his young life to use his talents to visually describe the events, surroundings and circumstances of his life. All of the training he had honed came together to lead him to his role in World War II. All of the events that he recorded through his sketches and paintbrushes brought his experiences to life in unique, heartfelt, timely, descriptive portrayals.

There is a sense of authenticity, age and history, that developed in my own heart. Dad's artwork needed to be shared, seen and felt as he observed and intended it to be. His amazing artwork and perspective of his war experiences is a treasure to be palpable within the viewers' sense of what occurred. I would hope that we would get a glimpse of how it must have felt to the soul of a young artist who chose to depict his experiences and communicate his observations through his own artistic eyes and expressions. The artwork needs to be presented as Dad created it, as he felt it. It has survived seventy-five years, through two long voyages packed with 8000 soldiers. It weathered a journey, shoved into a duffle bag, whilst my father slept in bombed out buildings with no roofs, and carted his belongings from location to location. It was carried across Europe and back to the safety of the United States.

After the war this depiction of his journey was moved several times, and spent many years stored away in cabinets in cigarette smoke filled rooms. Yet they are as alive today, as they were when they were created. They are depicted with memories, stories shared and emotions felt beyond words. This is the character I am hoping to portray in this daughter's telling of her father's memories and insightful artwork from years gone by.
As Dad put it tersely and efficiently, which was so much his character "What you see is what you get."
Life was not easy. War was not easy, or polished or computerized in those days. This is the art of the 1940's, done in the United States Army's training camps and encampments in war-torn Europe. It is the story that he harbored for most of his life before deciding that it deserved to be shared, as it told not only the story he experienced and painted, but also the story of the shaping of his life.

Richard E. Clark, World War II

Chapter 1

The Journey Begins
October, 2011

Through a Daughter's Eyes

Lynda Clark Porter, "Perspectives"

I peered down through the airplane window into the cottony layers of clouds. Although the sun was setting quickly, I took in the magnificent expanse of ocean below. I had a private glimpse of the deep blue water as it darkened into velvety night-time camouflage. We had departed some hours ago and had winged our way over the United States. The plane was now beyond the shores of Nova Scotia and flying silently in the trans-Atlantic flight to Amsterdam. My husband had organized a seventeen-day Autumn trip to Holland, Belgium and Tuscany to celebrate our 30th Wedding Anniversary.

As my eyes became heavy with sleep, I realized that I was thinking of my parents and their flight to Belgium fourteen years ago in 1997, at the time our family had been living happily for two years as ex-pats in the small village, Hoeilaart, on the outskirts of Brussels, Belgium. We had encouraged my parents to take advantage of the opportunity to visit us and see the sights of Europe. I was surprised and pleased when at age 79, my father had agreed to make the trip. I could recall only one or two other times that my father had flown, and I knew that it was not something that he relished. Boarding an airplane for an overseas flight was an adventure that I did not think he would readily accept. Never the less, my parents had booked flights to come for a two-week visit. Their two grandsons, Nathaniel and Benjamin, my husband Daniel, and I were very excited to share this experience with them.

I recalled that time fourteen years ago, just a few months before our family would leave Belgium in a transfer to London. I am not certain that I realized in 1997 what this trip would mean to my elderly father, a seventy-nine-year-old World War II veteran. I also did not realize the strain that this sort of travel would put on my parents at this time of their lives. Even at my age, a trip to Europe after a fourteen-year absence was daunting, therefore giving me pause at the magnitude it must have been for them. But as we glided through the clouds, and I peered into the setting sun and endless deep water below, I sensed what my father must have felt as he settled into his flight to Belgium. I imagined that he was thinking of the difference between this trip of a lifetime, and the first trip of a lifetime that he took in 1941. His view downward from the aircraft would have presented the memory of a ship far below, cloaked in darkness and parting the rolling waves as it moved forward meter by meter. The ship was loaded with 8000 soldiers deployed to Europe for their first tour of overseas combat duty. The contrast must have been overwhelming to Dad, some 55 years later. The visual image of the ship from a bird's eye view would have been overpowering, but also in every other aspect that an aged veteran would have viewed the most significant journey of his young life. I knew my father, and I knew that he would have pondered the symbolic notion of these two journeys. My father did not only see this viewpoint through the eyes of an aged veteran, but through the eyes of his soul as an artist.

My father was an artist. He was a self-made artist and fought for his self-expression from the early discovery of his talent and when he chose to follow art as his career. He worked hard to become an art teacher, and later an Art and Public Relations Administrator for a City School System. He met and eventually married his sweetheart Jessie, and they raised two daughters, my sister Becky and myself.

Richard Edward Clark, "Middleburg Home"

Richard Edward Clark, "Sherbourne Home and Farm"

Richard Edward Clark was born in Middleburg, in Southeastern Ohio. His father and mother, James R. and Clara Clark and Richard moved to Caldwell, a larger town a few miles North in Noble County, when he was a boy. Thirteen years later, his sister Thelma, whom he called "Tink" was born. As I recall, my grandfather often doodled and drew cartoon figures for us when we visited. Perhaps this was the spring well of my father's artistic talent. Their family survived the Great Depression, as did many families.

Richard E. Clark on Basketball Team, seated first row, 5th from left.

In later years, my father became known in his high school for his artistic talents which he eagerly used in year books and the school newspaper. He also demonstrated his lightning quick abilities in basketball. Dad was a very small man of about 5'6" and lucky to reach 108 lbs. He was proud of his high school basketball team who made it to State and won! He became an integral part of the team by weaving in and around the larger ball players to pass off the ball or score a basket. This little- known team continued on to win the State Basketball Championship that year. In the yearbook photos, he was last on the right, the smallest player among seeming giants.

SENIORS

RICHARD CLARK - "Clarkie": Classical course; Scholarship Test, Latin 1, 2: Basketball 3-4; French Club; Science Club; Latin Club, "Why the Chimes Rang"; "The Toy Shop"; "A Trip Through France"; "Officer 666"; Librarian 1-2-3-4; Class President 2-4; French Club President 3; Class Vice-President 1-3; Hi-Lite Staff

OFFICER 666

THE CAST

My father began to see that his artwork was important to him and integrated it more and more into his life. He also began to be seen as a leader and innovator, as he was elected President of his High School class two of three years, and Vice President the other year. He began to see that there was also a personal use for his artwork and began creating small watercolor paintings of homes in the area. He did paintings of relative's homes and gave them as gifts, perhaps an earlier version of 'pay it back'. I do not have a single memory of my grandparent's home that does not include an image of the painting hanging on the wall of the house where my dad was born. In later years, other paintings

Richard E. Clark, "Grain Elevators"

Richard E. Clark, "Sketch of Grain Elevators"

of my father's also took hold in my memories of my grandparent's home. One was of a large crooked brown and white splotched birch tree next to a bright blue stream. Another was a fantasy painting with large orange fish, horses and a flamingo standing on one leg. I cannot see in my mind's eye the overly warm living room, with its massively square brown gas heat stove and bright red leather hassock stool nestled beside it, without also seeing these wonderful creations that were my father's. My father later learned that he could also be commissioned for his home portraits and began to sell them as one means to put himself through college. Not only was he talented, but he was motivated and inventive.

Richard E. Clark, "Troy, Ohio Grain Elevators"

Richard E. Clark, "The Birch Tree"

Richard E. Clark, at Cincinnati Art Academy

Dad attended the Cincinnati Art Academy for two years after high school. Numerous examples from those years hung on our family room walls when I was in high school. He was very taken with perspective and would often paint the layers and angles of the houses and buildings which decorated the hills of the Mount Adams area of Cincinnati. He also developed an eye for imperfect subject matter, whether it was a charcoal depiction of an elderly male model, or a dilapidated building. He saw the good in it in terms of texture, interest, line, color, contrast and the need for it to be portrayed in an artist's medium. He saw that beauty is in the eye of the beholder and found beauty in unexpected places. In later years, when both my sister and I, as well as our fellow students, took our father's art class in high school, we learned well that a 'calendar' picture was not acceptable subject matter, but a broken-down shack was! He also developed an eye for the odd angles and roof lines of grain elevators. These depictions hung in our living room for years. Our home was always a sort of gallery for Dad's art. After Becky and I began producing our own art in high school and college, Mom and Dad hung dark cork squares on the end wall on our family room. This evolved into an extended gallery of all three artists' work. The end wall in the living room soon became a display of our artwork as well. In later years when my mother began to paint, her paintings also graced the walls of our dining room. It was evident that creating art was revered in our home.

After earning the funds to continue his education, my father was admitted to Ohio State University. He lived in The Stadium Club, a dormitory which was located under the Ohio State Stadium. He also became active in Delta Phi Delta, Words, Triskelion (yearbook), Pen and Brush Club of The Stadium Club, and the YMCA. He was a born leader and saw a need to bring people together, organize and do good for a cause. Dad continued to develop his creative spirit, his artistic eye and his organizational skills.

The photo on this page shows my father standing behind the easel with an illustration of the "Cannon over Cincinnati." The original watercolor is depicted on the opposite page. Subsequent illustrations depict scenes in and around Cincinnati that my father painted while attending the Cincinnati Art Academy.

Richard E. Clark, "Cannon over Cincinnati"

Richard E. Clark, "Cincinnati Bridge"

Richard E. Clark, "Church with Bridge"

Richard E. Clark, "Cincinnati Skyline"

Richard E. Clark, "Above the City"

Richard E. Clark, "Market Day"

Richard E. Clark, "Life"

Richard E. Clark, "Above the River"

Richard E. Clark, "Church in the City"

Richard E. Clark, "The Plow Horses"

Richard E. Clark, "Building on the Hill"

Richard E. Clark, "The Arboretum and Mount Adams"

Richard E. Clark, "Church with Bridge"

Richard E. Clark, "Watching Over All"

Richard E. Clark, “Man with a Letter”

Richard E. Clark, "Gentleman with a Cane"

Richard E. Clark, "The Sanctuary"

Richard E. Clark,
"The Cross and the Church"

Richard E. Clark, "Up the Hill"

Richard E. Clark, "Woman in Red Dress"

Richard E. Clark,
"Passage"

Richard E. Clark, "Back Street"

Richard E. Clark, "Work and Play"

Years later, after my sister, Becky, and I had also become art teachers, my parents began to winter as 'snowbirds' in Tucson, Arizona. They began with a week in the winter; then two weeks, until they eventually bought a small motor home in a retirement community on the Southeast side of Tucson. My husband and I first visited them in Arizona, soon after we were married in the early 1980's. We had often heard of the azure blue skies, shadowed violet mountains, brilliant sunshine and stoic Saguaros, via regular phone conversations with my parents. Upon our first visit, we soon realized that we were also desert loving snowbirds, as we still are today.

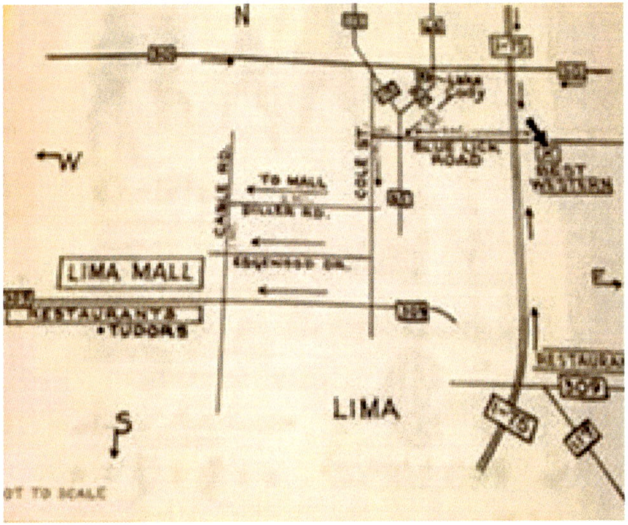

Richard E. Clark, "Map"

When we visited Arizona, Dad would deliver his infamous, itinerary and lecture about the streets of Tucson. Red and black flair tipped markers were used to draw maps on yellow lined legal pads. He used these to illustrate and direct us to the North and South and East and West streets from Mountain range to Mountain range; one way-streets, and most importantly, the location of convenient gas stations en-route to the airport. The presentation was always delivered with a waving arm and a pointed index finger. As years passed, more and more maps were drawn, for family events, such as weddings, picnics, vacations to Arizona, and even maps to Metz France. It became a family chuckle and endearment to be presented with yet another Papa Clark Map. I remember constructing a map for both sons when they went off to respective colleges, drawn in the perfected Papa Clark style. As Dad aged, he developed palsy in his right hand that ironically, and sadly made it difficult for him to hold a pen or a paintbrush. His maps were still drawn and presented, but with a very shaky appearance, as if he had used a spirograph pen. My husband also learned that the Clark

"The Richard E. Clark Wall of Honors"

Map fascination had been passed on to his wife, as I became somewhat obsessive about making my own maps, or at least highlighting routes yon and hither on maps of European destinations. Unfortunately, in my husband's opinion, I did not inherit the ability to read a map or decipher north from south, east from west or right from left while attempting to navigate from a map.

Although, we often chuckled about Dad's map lectures, it did not occur to me until years later that there was a significant reason that my father was obsessed with mapmaking and mapmaking skills. My father never wanted to discuss much about The War. We knew that he had served in Europe during World War II. There was a rusty iron sword and a rugged helmet of unknown origin in the basement. There was also talk of an old pistol that was hidden somewhere. We observed pins and medals mounted on blue velvet in a simple black frame on the family room wall. It was displayed amongst the array of Lions Club, civic and community club plaques and awards. How many times had I walked past them, dusted them for my mother, or straightened them without really looking at them? We took for granted the display that had always been there.

My Dad was very active in the local 95th Division Veteran organization. To me, that meant meetings and dinners he attended now and then, and speeches delivered. A gray-haired fellow and his wife, from a neighboring small town would sometimes stop by and visit for hours, talking about 95th Division activities and events. Dad designed flyers and invitations for the local reunions and meetings. Much later I recognized that these were renditions of drawings he had created long ago.

We did not have a great deal of perspective about World War II. In the 1990's the 95th Division Association decided to request veteran's interviews about their wartime experiences. Although difficult, my father must have decided that it was time to revisit some of those memories. He asked my brother in law, Jerry, if he might video tape a ten-minute presentation of Dad talking about the war as he discussed some of the art work he did during the war. When we visited, my father brought out a large, tattered scrapbook. In it, there were newspaper headlines, articles, photos and memorabilia from that time of his life. I sensed the air laden with a heavy significance, as he opened the worn cover and began to reveal its contents. He shared that his mother and sister had saved many articles and photos to bind together into a scrapbook. Never before had he acknowledged its existence, presence or the monumental influence of the war in the shaping of his and my mother's lives. I remember that my sons, Nathaniel and Benjamin were present and I realized the significance for them as well. His voice was poignant and memorable as he revealed unspoken, difficult memories and events. On that day I learned that my father had been recognized for his artistic skills in the war. He became a Master Sergeant in the Division Artillery Headquarters, as Operations Sergeant. Eventually he was assigned to scout ahead and set up headquarters for his Commander; but he also worked with mapping and plotting routes and artillery cannon projection within his Battalion. This, in and of itself, made World War II very real for me in a flash of a moment, in a way that I could not have previously recognized or understood.

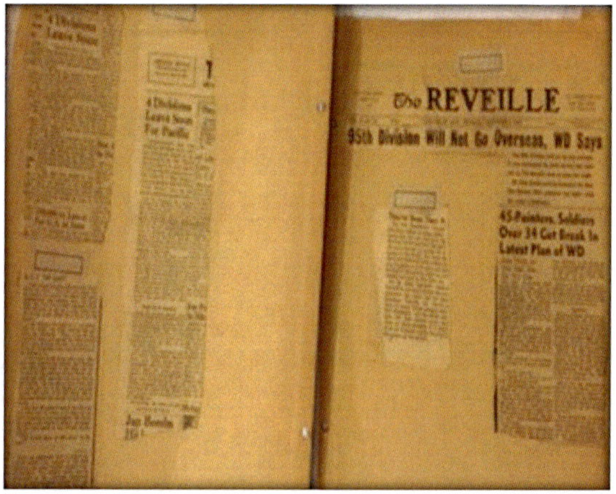

Richard E. Clark "Scrapbook"

So, as it goes, The Mapmaker had made maps for a very significant cause. Those maps went on to help bring success for his Division's maneuvers and endeavors during World War II. The Mapmaker never stopped making maps, and he never stopped being the Patriot whose spirit was deeply influenced by what he saw, what he experienced and what he lived during the war years.

My parents loved to go to a favorite pizza parlor in Tucson with, oddly enough, a backdrop of organ music. When they suggested that we go there for dinner on that first visit back in the 1980's, we had no idea what to expect. We walked into a building that resembled a large meeting hall. It had long rows of benched tables much like a school cafeteria. After we had placed our order, taken our seats, and had a few sips of our sodas, the lights dimmed, and a prominent display of organ pipes became apparent on a stage at the front of the room. The organ pipes appeared larger than life against a purple velvet curtain. The room darkened and spotlights were directed at the pipes in pure Hollywood style. Red white and blue light beams began to filter in with violet hues. The first bars of The Star-Spangled Banner began. My husband and I looked at each other, wondering how this could be happening in a pizza restaurant. But sure enough, every person in the room stood up at attention, with hand over heart and fervently sang The Star-Spangled Banner. We were embarrassed at first, as we were in a room that was not very full, of not very young people. Everyone there appeared to be of the older generation, and every one of them had their hands on their hearts. My mother and father were standing straight and tall singing proudly and loudly; particularly my mother who had an operatic voice. I found that my hand went to my heart, and I stood and sang along as I had in high school choir days, full of emotion and love for my country. We proceeded onto other songs such as America the Beautiful, The Battle Hymn of the Republic, and God Bless America and I could not help but blot at the tears forming in my eyes.

I caught a glimpse that day of true patriots, with patriot dreams and patriot hardships, that I had never known. I saw in my heart, my parents as they stood for The Star-Spangled Banner with hands on hearts, at high school football games, Lions Club Events, Parades, Graduations, civic events and many others. My mother always embarrassed me with her voice much louder than everyone else, but she was singing out for her country and the freedom that had been defended on foreign soil. She, too, had worked with other women during the war to support the effort. I saw my parents marching in community parades and International Lions Club parades, wearing red and white and pulsing their patriotism through every living breath and heartbeat within their beings. They felt the love of country as their predecessors had, and they knew the meaning of defending the rights of others, freedom, and of coming to the aid of others. Their lives were a day by day example of giving and helping neighbors, friends and community. They lived a life they had fashioned from their own beliefs, contributions and life experiences as members of "The Greatest Generation." They were "The Greatest Generation." After realizing this I did not feel embarrassed to place my hand on my heart and fervently sing the anthems of my country. I had finally seen the hearts of The Mapmaker and his wife.... My mother and father.

Richard E. and Jessie Clark

RICHARD EDWARD CLARK

His Early Years

Richard E and sister, Thelma "Tink"

Caldwell Home

James R, father, holding Thelma while Richard stands next to him.

Middleburg Home Neighborhood

Chapter 2

The War Years

Decades later, as I began to piece together the story of my Dad's life, I came across many worn manila file folders with Dad's hand-written block letters on the tabs. While sorting through these files, I came across the goldmines of his life, and the treasures of mine. I found speeches, words to live by, favorite quotes, poems, philosophies, notes and histories. All of these were the bits and pieces I had searched for to honor my father. Within these files, I found his point and peck style of typed outlines and notes for his speech to be delivered on November 4, 2001 at ArtSpace of Lima Ohio. ArtSpace offered to host an expose of his World War II art work, and the opportunity to deliver a talk about his experiences and artistic expressions. As I read through his notes and talking points, and several of his other speeches, I realized that this was the perfect format for presenting these years of his life. What could be better than his own writings, descriptions, voice, memories and emotions? Below are excerpts from Richard E. Clark's Exhibit and Presentation to a gathering of his family, former coworkers, friends, associates, teachers, former staff members, Lions Club buddies, artists and friends from the Local 95th Division Association. Excerpts from other speeches, my commentary, conversations and his writings are included.

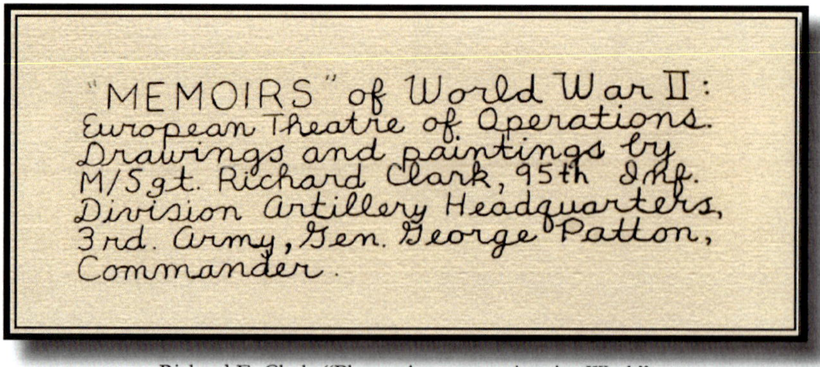

Richard E. Clark, "Plaque Accompanying Art Work"

Highlights and Summary of Richard E. Clark's World War II Service

Written and presented by Richard E. Clark
November 4, 2001 Art Space Lima, Ohio

"The Odyssey of the 95th Victory Division," 3rd Army, General George Patton, Commander Drawings and Paintings by then Master Sergeant Richard Clark, Operations Sergeant

The Voice of My Father, Richard E. Clark

Richard E. Clark speaking at ArtSpace Lima

"It's a pleasure for me to be invited by ArtSpace Lima to display some of my art work and tell my story of The Odyssey of the 95th Division. If you go to a high school or college reunion, everyone is always saying "It seems like old home week," and that is the way I feel today. Former students, teachers, my boss for eleven years at the educational center, several people with whom I was associated with on committees or boards, the original Arts Council, Citizens School Advisory Committee, Senior Citizens Board, School Board, United Way members, special guests and family…

I'm sure all of us are with heavy hearts today in view of the dastardly bombing of the World Trade Center in New York City and the Pentagon Building on September 11th. I, like everyone else in our country and the world, was in total shock and disbelief that this would happen in our country.

For the last several weeks, I've felt that time stopped, stuck at the moment, that our past was behind us and seemed forgotten and insignificant at this time. Before our very eyes we witnessed an act of war taking place in our country, which has triggered another war of global involvement. And the great wave of support for our nation's leaders, and the deaths and heroic deeds of the firefighters, police, doctors and nurses, and volunteers make these people veterans of this war. All of these things make you recall things from your past.

It is very difficult for me today to get my mind on the past. But as our President said, clergymen said, and even the relatives of those killed have bravely said "Life must go on," facing what lies ahead, but also remembering the past, as this horrendous event will be in our past, and we will remember the lives lost and courageous efforts of the living and pay tribute to them in the future… as another date to be remembered in the history of our nation.

And so it is today. Next week we will remember past events and honor the veterans of past wars who have served our country in times of other wars. Armistice Day is November 11, which has been changed to Veterans Day. Armistice Day was established to pay homage to the men and women whose lives were sacrificed in protecting our country and democratic nations. It goes down in history commemorating the cessation of World War I. Today the veterans of all wars are honored on Veterans Day.

Our country became involved in World War II after the bombing of Pearl Harbor on December 7, 1941. I can well remember that date and incident. I was a student at Ohio State University. My hometown was Caldwell, Ohio, half way between Cambridge and Marietta. I often hitchhiked home, but this time I had caught a ride with my cousin who was also in Columbus and was on my way back to Ohio State University, when President Roosevelt made a radio announcement. These became very famous words "A date that will live in infamy." Pearl Harbor had been attacked. There was complete support for President Roosevelt. Thousands of men and women left their jobs and enlisted in the armed services to defend their country. Congressmen left Congress to enlist. Movie stars, Clark Gable and Jimmy Stewart quit and enlisted. Students at Ohio State dropped out of school to enlist. I even had thoughts of it and went to the registration center in Columbus to enquire about enlisting. There was a line blocks long for paratroopers. They were not interested in me. I was afraid of heights and I weighed 108 pounds and figured I would not even weigh enough to float down to the ground! I went back to my dormitory to wait until I was drafted.

I'm not going to talk about Lionism today, which is unusual, but about my own profession in art education and how I used art while serving in the army during World War II. I'm sure some of you served in a branch of the service during some war and probably had similar experiences, so my talk may bring back some memories.

I'd like to give just a little about my personal background. I'm originally from Noble County in the hills of south eastern Ohio, Appalachia. We were poor folks. I graduated from Caldwell High School, the county seat, 1800 population. I attended school during the depression years, as many of you will remember those days. My dad was out of work for two years. I got a job delivering papers, The Columbus Dispatch, for the four years I was in high school. Believe it or not, the money from my paper route and a few odd jobs my dad had, kept us eating - bread was five cents a loaf.

I didn't get any training in high school. I had a knack for drawing, copied some magazine pictures and painted some signs in town. But my family didn't have any money to send me to college. I landed a job working at the five and dime store, which some of you will also remember, at $1.50 a day, $9.00 a week, and I saved enough to start college the following year.

I studied two years at the Cincinnati Art Academy and worked part time at a YMCA. I ran out of money after

Richard E. Clark, "Designs of Emblems"

two years. I got a job in Columbus at an athletic uniform company designing school emblems and lettering uniforms for high school, college and professional athletic teams (Cleveland Browns, Cincinnati Reds). While in Columbus I enquired at Ohio State about going to school there, and I got a scholarship for students who couldn't afford to go to college (we were indigents). We worked for our room and board. I lived in dormitories built under the Ohio State football stadium. There were three clubs there. I was in the Stadium Club. I cleaned rooms, served meals in the cafeteria, washed dishes, etc. I heard all the cheering and noise of the football games above, but I had to listen to them on the radio as I didn't have the money to attend the games.

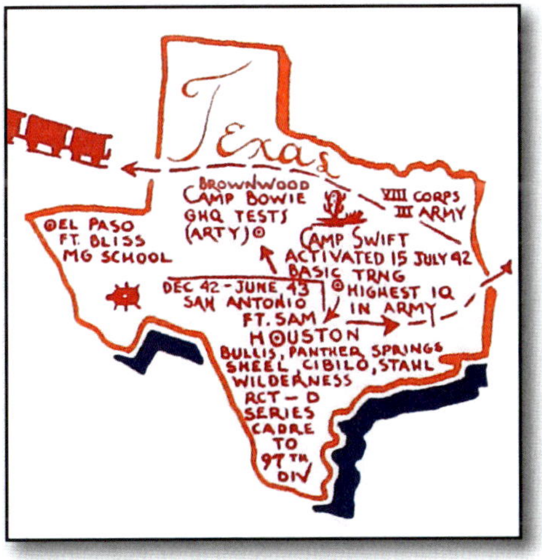

I graduated in 1942 with a degree in Art Education. I was deferred until I graduated in June of 1942 and was immediately drafted into the army, so my teaching career was put on hold for the next four years. I was sent to the induction center in Indianapolis, and then to a new camp in Texas, Camp Swift. I started basic training as a member of the newly organized 95th Infantry Division and was placed in the Artillery. I started out as a buck private, but was promoted to Sergeant in a month, and to Master Sergeant in another month. The reason was it was a new division. But I had something going for me. I had two years of ROTC Training in Artillery at Ohio State. I knew a lot about it, but I listened. I was lucky to be placed in the Artillery Division.

We began basic training which included the physical phase of marching, drill sergeants, rifle training, (I fell backwards several yards when I first fired a rifle); and infiltration training- crawling under barbed wire a foot over the ground. When the basic training was complete I was told that I would be Operations Technical Sergeant and that I would be instructing Artillery. I skipped the levels of Corporal and Sergeant. I began at 159th Field Artillery sixty miles South of Austin, TX. I conducted technical instruction, to three Infantry Regiments, three Artillery Battalions, one Light Artillery Battalion and one Heavy Artillery Battalion. I instructed guns, cannons with tall narrow barrels, to be pointed at any angle to reach plotted targets. They could be tilted up at any angle, and sideways. They could reach from short distances up to six miles. There were scouts sent up ahead of the front-line Infantry, and Artillery behind the regiment front line. There were three companies of guns, each company had four guns. The Light Battalion had thirty-six guns and the Heavy Artillery had twelve

"Field Artillery in Action Fort Sam Houston"

guns. There would be thirteen small artillery battalions that were behind the Artillery, three companies with forty-eight guns. Along with the fighter planes, this was a lot of power.

As Technical Sergeant, it was my job in the Battalion to figure it all out. I had to figure out how to change angles to fire several yards up to six miles. There were scouts ahead out in the field assigned to the Infantry who radioed back where the enemy was. Maps were plotted to figure the distance and where to fire. My job was to instruct all of this. When all of this was called in, it was a whale of a lot of power. Our job was to soften up the enemy so that the Infantry could do their job. "If you didn't kill them, they would kill you."
Two weeks later, I was told to report to Division Headquarter at Fort Sam Houston in San Antonio, Texas. I packed my duffel bag and headed to Division Headquarters. The Operations Sergeant had been "busted'- fired. He was the Operations Master Sergeant of the Headquarters, with twenty Officers; Lieutenants, Lieutenant Colonel, Major and General. I was to be the new Operations Sergeant. I would not be instructing anymore. I became Master Operations Sergeant, the highest rank as an enlisted soldier. I was in charge of Artillery Operations in Headquarters. As Operations Sergeant, I was in charge of making maps of enemy territory, plotting firing data, and running the operations of the command center.

Richard E. Clark, "The Big Picture"

My job was under the General at the Artillery Headquarters in handling plans and training for the Artillery Battalions, making maps from photographs taken of the area by personnel on small two seat planes who flew around over enemy territory taking photos. 8"x10" detailed enlargements were made of these photos. I made a mosaic from the photos, and maps from the mosaics showing landmarks, roads, rivers, forests, towns etc. for the Artillery Battalion's cannons. I taped the photos on a big wall in order to create the pieces of the puzzle that made the mosaics. This was the Big Picture of the Whole Front. I gave briefings to the Officers

Richard E. Clark, "The Big Picture"

on the location of enemies, and of the General's troops in battle. Overlays were given to all of the Officers. This was all done on gel mimeographs and the equipment of that era. It was not an easy process. Later, it was also my job to find a place for my men to sleep, set up tents, etc. and set up the headquarters. (An excerpt from Armistice Day Speech to Lima Optimist Club.)

After my time at the Headquarters, we then were sent to the swamps of Louisiana for training in case of duty in the South Pacific. It was very, humid, hot and rainy with lots of water. Next we were sent to the Mohave Desert outside of Yuma and in Eastern California, for training in case we were sent to the African Desert to fight General Rommel's army. It was very hot and dry. One day I sat and watched as a tick burrowed into my arm. I noticed that it burrowed in a clock wise direction. I discovered that if I were to turn it in a counter clock wise direction, I could lift it right out. I began to spread this discovery amongst the soldiers and it provided relief from the thousands and thousands of ticks that made our lives miserable in the Mohave Desert.

I received a pass and a few of us were able to go to Palm Springs and Hollywood. We went to the most famous Canteen that was set up for the soldiers. It was a very large room, almost the size of a basketball arena. There was a band at the back playing the hits like" White Cliffs of Dover"' and refreshment tables. I was able to get the autographs of three famous ladies, Hedy La Mar, Donna Reed, and Susan Hayworth. I also was able to dance with them for all of about two minutes before the next soldier stepped in for his turn!

Marlene Dietrich and Rita Hayworth serve food to soldiers at the Hollywood Canteen in 1942

Lynda C. Porter,
" Train Station Lima Ohio"

Richard E. Clark
"Black Water Falls WVMA" West Virginia

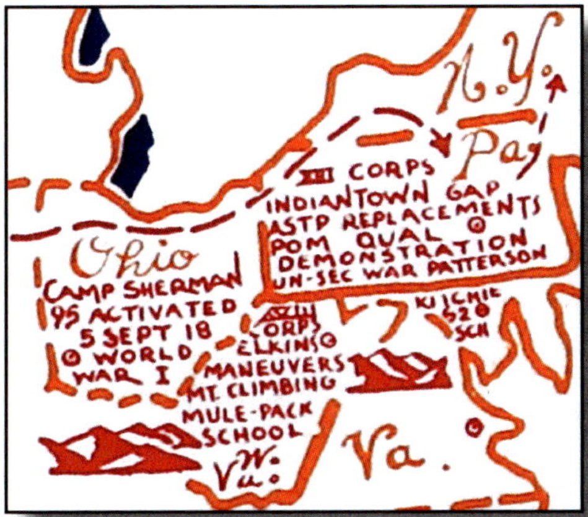

After this, we were sent by train to Indian Town Gap, Pennsylvania. I distinctly recall pulling into the station in the town of Lima, Ohio, where the USO women were passing out coffee and donuts through the train windows. They were very appreciative, waving flags and cheering us on. It is really something that many years later, I ended up spending many years of my life here in Lima with Jessie and our family. When we arrived at Indian Town Gap Embarkation Center, we thought we were there to wait for our next call of duty to go overseas to Europe. However, we were then sent to Black Water Falls, West Virginia for mountain training. While I was stationed there, my mother and girlfriend at the time, came to visit me.

We finally received our orders to move on to Boston Harbor, where we would disembark on our voyage overseas. I remained in this same rank and position for the rest of my service. I guess if I were to title my talk, it would be "The Odyssey of the 95th Infantry Division" the unit I served in.

I am a veteran of World War II, as were millions of others. I am not here to brag about my experiences, I don't think any veteran will brag about their experiences. War is a terrible thing. I was fortunate, lucky. I came home. Thousands didn't. Thousands were wounded, maimed for life, physically and mentally. Thousands were taken prisoner or missing in action (John McCain is a good example.) Thousands died in battle and were buried overseas. I have some figures if you would like to hear them.

Figures of Veterans Who Died In Action In Various Wars:

Civil War	(1861-1865)	
	364 Union Soldiers, 133,821 Confederate Soldiers	
Total:		498,132
World War I	(1914-1945)	116,516
World War II	(1941-1945)	405,399
Korean War	(1950-1953)	54,246
Viet Nam War	(1962-1975)	58,135
Persian Gulf War	(1990-1991)	299

All veterans had experiences, all different – some good, some bad… some so horrible that veterans wouldn't talk about them. Recently Veterans are telling their story. On Memorial Day National Geographic had a program on TV about the bombing of Pearl Harbor, narrated by Tom Brokaw. Veterans on the program said they had never told their story before, but they thought it was time, and they did.

Last Veterans Day the Lima News printed a large magazine type edition that told the story of many Lima veterans, stories from all branches of the service—Army, Air Force, Marines, Navy … Stories from Pilots, Infantry men, Tank soldiers, Cooks, Supply Sergeants etc. On this day of remembrance, I have had some experiences that I feel I can talk about that relates to Veterans Day. In order to do this, I have to be very personal because the experiences were mine and affected me and my life very much.

I was a soldier serving in the European Theater of Operations. While I was there, I made several paintings, water colors. They are 56 years old. They have been in a portfolio laying on a shelf for many years. I never showed them to anyone except my family and a few friends, and three or four organizations in town.

Richard E. Clark, "Rincon Country"

Jessie and I spend our winter months in Tucson Arizona. We have been going there for 20 years. I teach art classes there in our community. I took my paintings there last winter because I wanted to show them to my art classes; display them at the community arts and crafts show, and also at an art gallery. There were a number of veterans in the park who were touched by my paintings; as well as the Editor of Arizona Magazine who saw them. When I talked with Tina, Director of ArtSpace, at Lions Club and showed her a sample of the paintings, she enquired if I might be interested in presenting them with a personal talk for Veteran's Day at ArtSpace Gallery. That is the reason I am here today. I plan to take them with me, so I didn't buy 25 frames. They are in their original form; "what you see is what you get."

Chapter 3

The Sketches

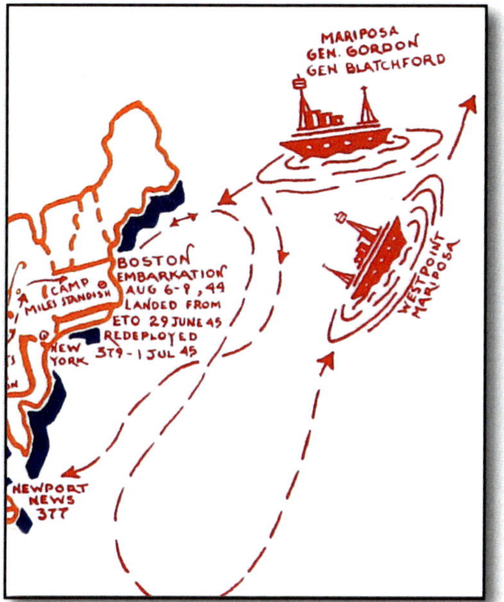

Richard E. Clark, "Boston Harbor Embarkation"

When we arrived in Boston for Embarkation, I had a few moments to run out and purchase a couple of sketching pencils, a pad of paper and a small set of watercolors which I threw in my back pack. I was hoping that I might have a chance to record a bit of my experiences through art. The Division Set Sail from Boston Harbor on the U.S.S. Westpoint on August 6-7, 1944. We hit submarines up our shoulders 200 miles out and had to divert several hundred miles.

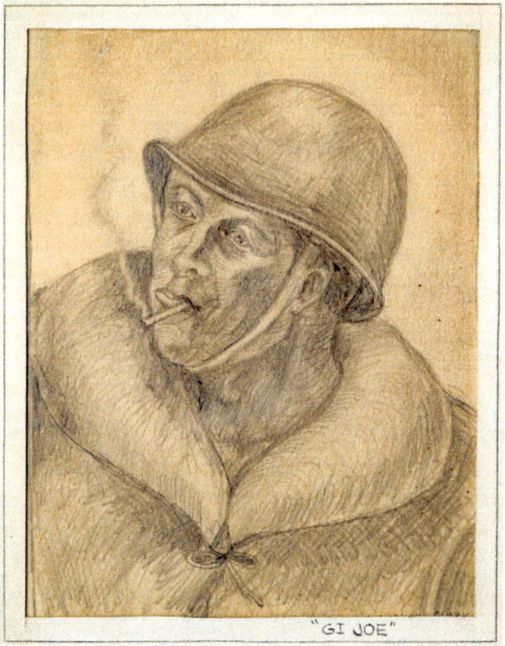

Richard E. Clark, " G.I. Joe"

Richard E. Clark, “On the Train”
En route from the Mohave Desert in
California to Indiantown Gap, PA

Richard E. Clark, “Embarkation - Aug. 9-44”
Adjusting equipment and walking up the gang plank to board the troop transport ship to go overseas.

Richard E. Clark, “Quarters”
The sleeping quarters were bunk beds from floor to ceiling. The soldier and all his belongings had to fit into the bed.

Richard E. Clark, “““Chow”
The soldiers going through the chow line with their mess kits while trying to keep their balance and weave through the ship and manage the rough seas.

Richard E. Clark, “Private Bath”
The private baths had to accommodate about twenty soldiers.

Richard E. Clark, “Rest on Deck”
On board were 8,000 men and finding a place to sit or lie down was a problem.

Richard E. Clark, "Emergency Drill - 8000 On Board"
An emergency drill was held each day with instructions on how to use the life jacket ad jump overboard in case of being torpedoed.

Richard E. Clark, "Rumor - Where we're Headed"
Although we knew they were crossing the Atlantic Ocean, rumors were spread as to where we were going (Africa, Italy, France or Emgland).

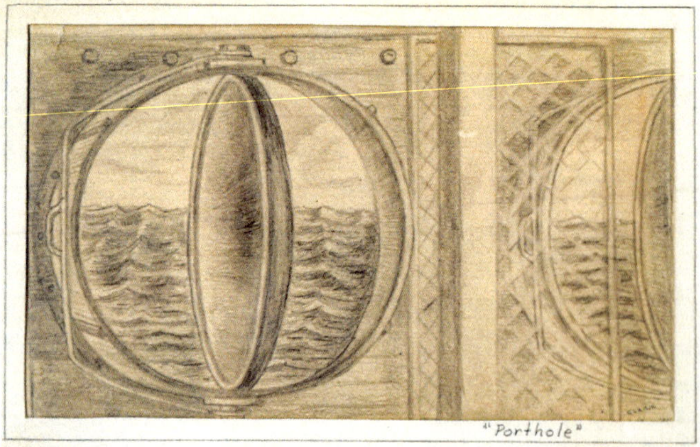

Richard E. Clark, "Portholes"
When the portholes were open we could see the ocean.

Richard E. Clark, "Mass by the Mast"
The military chaplain conducted daily services
for those wishing to attend.

Richard E. Clark, "Close Portholes"
Portholes had to be closed at dusk, as a light from one
could be seen for many miles.

Chapter 4

Rumors No More

Arriving in England

Richard E. Clark, "Landing in Liverpool"

"We landed in Liverpool, England August 14-15, 1944.

The Division had a brief stay at Camp Barton Stacey near Winchester, England. We were stationed there with the Artillery at Gloucester. We were waiting about five weeks for our assignments and dates to cross the English Channel. While we were stationed here, the troops lived in Quanset huts.

Our next station was at Dorchester, England. Metal, arched shaped huts were the living quarters for troops while in England. English buildings and garden patches are in the background.

Richard E. Clark, "Quanset hut, Dorchester, England"

Richard E. Clark
"Pub Whitchurch, England"

This is a typical pub in a small town, with a bar, tables, pool table, soldiers and women, "Half and Half" warm beer or ale was served, with an Englishman playing favorite pastime of darts.

Richard E. Clark
"English Pub - Countryside"

Residents outside the town often opened their homes and served soldiers in their living room. In this scene a rather staid Englishman was being served ale by an elderly Englishwoman. The Englishman wasn't a conversationalist by any means.

Richard E. Clark, "Thatched Roof Home - England"

These houses dotted the countryside. The roof was made of straw and hay. Water runs off the roof keeping the house dry.

Richard E. Clark, "Salisbury Cathedral, England"

A beautiful cathedral with a single delicate spire in a beautiful setting. I was able to take some time to go view it, and compose a painting."

Richard E. Clark,
"Piccadilly Circus"

The night life of downtown London is depicted. Where there are soldiers, there are women (and visa-versa). My Captain offered to go into London town with me. Of course, I did not go in.

Chapter 5

D-Day + 89

Entering France

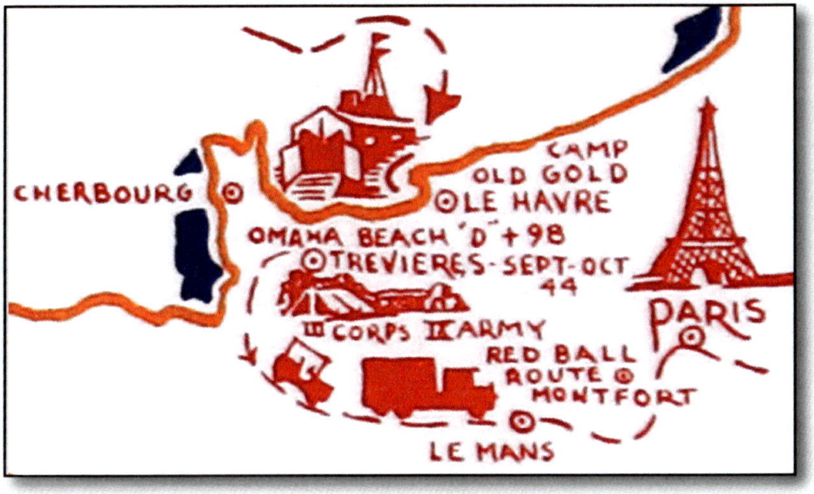

Richard E. Clark, "Entering France and The Red Ball Route"

"We could hear the German rockets over the coast of France. The propellers sounded like a motor coming; but they were erratic and not accurate. They were used as psychological warfare, and frightened the people.

The Division embarked via LST flat-bottom square PT boats, crossing the English Channel, landing at Omaha Beach, France on D-Day + 89. The boats let down the steel doors and we marched right out onto the land. On D Day we had lost 12,000 men in the invasion. This was the great invasion of Europe, with Air force, Navy and Armies. But a foothold in France was gained and the Germans were pushed back. The frontlines were half way between Germany and France.

The Division camped at Trevieres, west of Paris in preparation for deployment to the front lines and active duty. While there, a cadre was formed from Division Artillery to setup and operate the headquarters for 'THE RED BALL ROUTE,' a truck route carrying supplies for the French Coast distribution center to the front lines. I, Sergeant Clark, was Operations Sergeant for this assignment, located at Montfort, France, south of Paris.

It was my job to set up Headquarters. A rich French family had donated the use or their house for us, which was very gracious of them.

95TH DIVISION TASK FORCE RED BALL ROUTE HEADQUARTERS, MONFORT, FRANCE

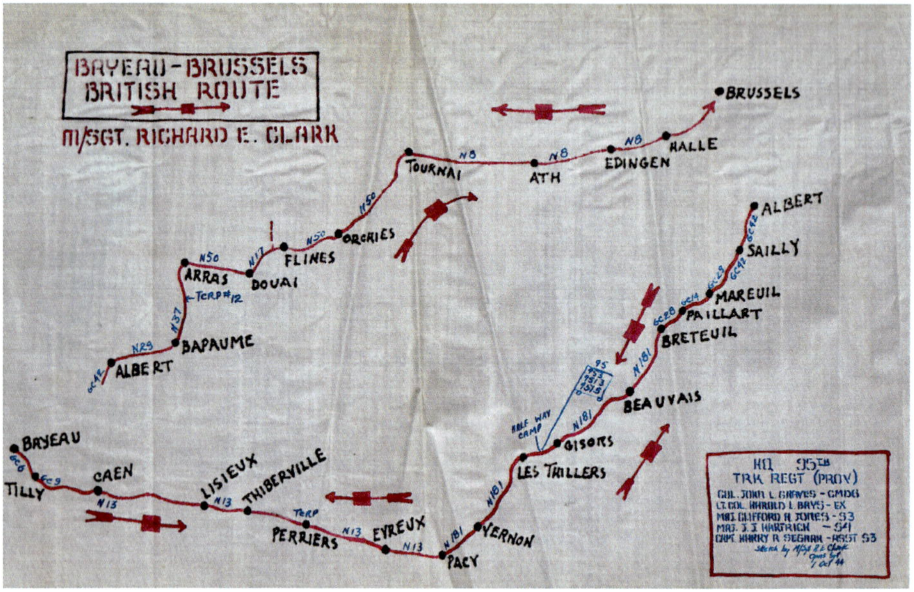

Richard E. Clark, Cloth Map of "Bayeau - Brussels - British Route"

The 95th Division Artillery set up and manned the Headquarters for the Red Ball Route, a truck route that carried supplies from the coast of France to the front lines. Today on our interstate highways semi-trucks can be seen with a red ball on their rigs, named after the Red Ball route in Europe during World War II.

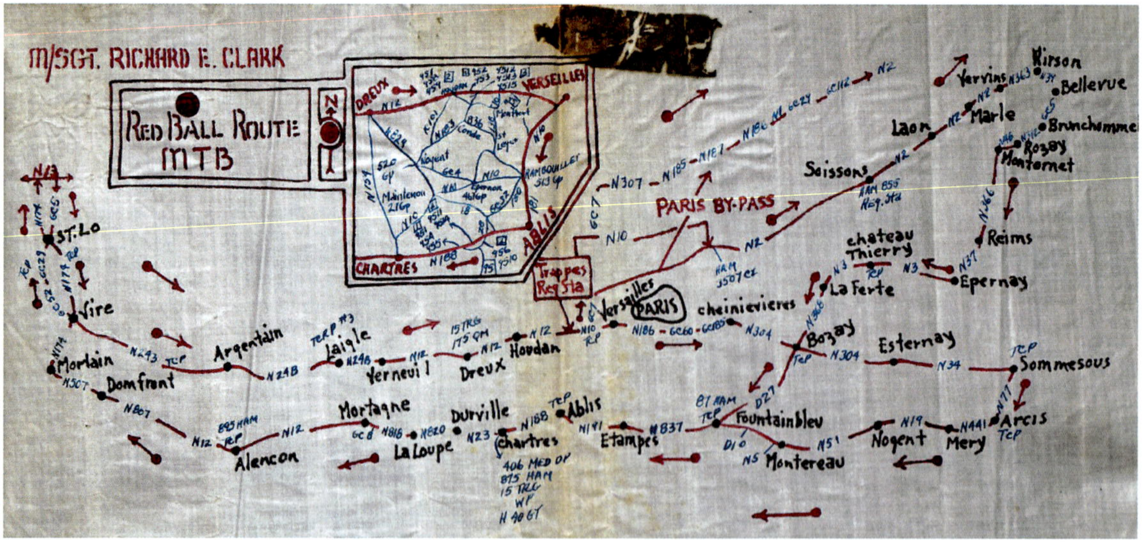

Richard E. Clark, Cloth map of "Red Ball Route MTB"

Richard E. Clark, "95th Division Task Force Red Ball Route Head Quarters Monfort, France"

Richard E. Clark, "Monfort Church, France"

Monfort Church, Monfort, France – a typical village in the middle of the square with businesses around the square, and cobblestone streets.

We were just waiting for our next move, so I was able to wander into the town of Monfort. I painted the church there and was pretending that I was painting Sacre Couer in Paris. When it was dark, or black out hours, I learned to put my coat over my head to create a sort of tent; use a torch for light; and place the water I needed to paint with watercolors in my upside-down helmet. This came in handy later in times where we stayed in open top floors under open skies.

We were also given a pass to go to Paris. We went to the Moulin Rouge- (although I did not look!)"

Restaurant Obligado (pres l'Arc Triomphe)

Eiffel Tower

Chapter 6

The Iron Men

The Bravest of the Brave

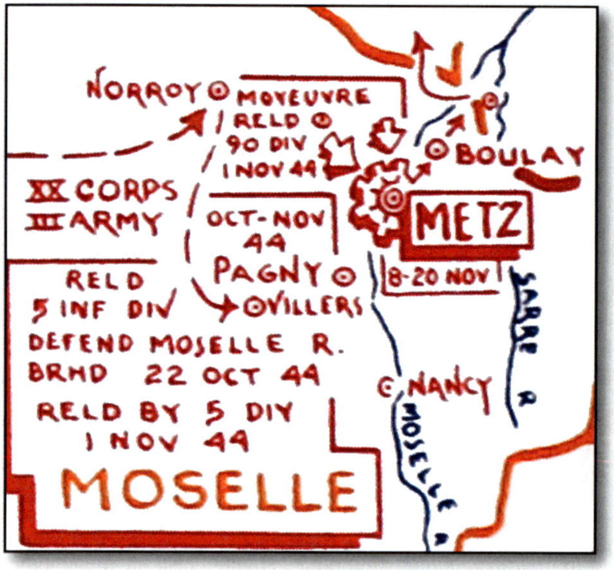

"The 95th Division became a part of the 3rd Army under the command of General George Patton. The Division was ordered by General Patton to move near the front lines, replacing the 90th Infantry Division west of the City of Metz. This was a key War Battle and they needed to take a few towns, including the city of Nancy. I had plotted the city of Nancy. There was a bridge there which honored the 35th Division, where another soldier from Lima was buried. Years later, his mother placed a plaque in his and the 35th's honor at the base of the bridge.

It was our job to take the city of Metz. The city of Metz, with a population of 300,000 was a cultural and industrial city located on the Moselle River and was fortified with the Marginot Line, which had been constructed by the French during World War I to protect France from any conflict with the German Empire. The Marginot Line was a big Engineering Masterpiece. It was built from 1920-1939 and ran approximately 475 miles from the south of Metz to near the Belgian border. It contained pillboxes and trenches to protect France from the German army. The Germans by passed it totally and went through Northern Holland and Belgium in six weeks. In fourteen days, they all surrounded Metz and set up government, mayors in all of these countries.

A fold out postcard from the city of Metz, France owned by Richard E. Clark

The masterpiece was Metz. It was surrounded with forts, pill boxes and bunkers. The bunkers were two stories, above and below ground structures, which gave Metz the name of "The Underground City." Above ground the structure "Y" was a circular structure resembling a silo with a round dome with openings for machine gun fire. The walls were two to three feet thick. Twenty feet underground was virtually an underground city with a network of tunnels which stretched for five miles. The tunnels, which branched out in different directions, contained barracks housing 2000 soldiers, classrooms for military instruction, a hospital/dentist office, dining room, and exercise room, trollies to transport the soldiers to and from the above ground structure, entrance and escape exits, through which supplies could be brought. After Metz was captured by the Germans it became the training center for fanatic German officers.

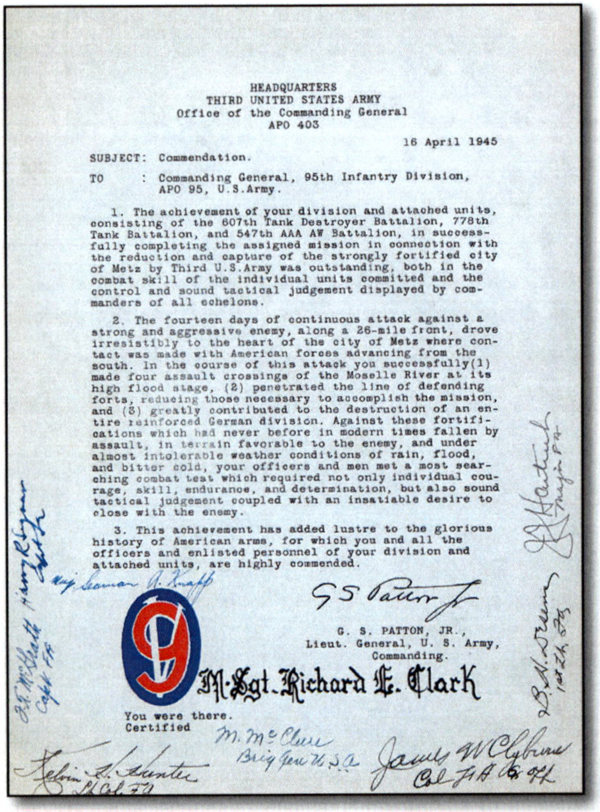

HEADQUARTERS
THIRD UNITED STATES ARMY
Office of the Commanding General
APO 403

16 April 1945

SUBJECT: Commendation.

TO : Commanding General, 95th Infantry Division, APO 95, U.S.Army.

1. The achievement of your division and attached units, consisting of the 607th Tank Destroyer Battalion, 778th Tank Battalion, and 547th AAA AW Battalion, in successfully completing the assigned mission in connection with the reduction and capture of the strongly fortified city of Metz by Third U.S.Army was outstanding, both in the combat skill of the individual units committed and the control and sound tactical judgement displayed by commanders of all echelons.

2. The fourteen days of continuous attack against a strong and aggressive enemy, along a 26-mile front, drove irresistibly to the heart of the city of Metz where contact was made with American forces advancing from the south. In the course of this attack you successfully(1) made four assault crossings of the Moselle River at its high flood stage, (2) penetrated the line of defending forts, reducing those necessary to accomplish the mission, and (3) greatly contributed to the destruction of an entire reinforced German division. Against these fortifications which had never before in modern times fallen by assault, in terrain favorable to the enemy, and under almost intolerable weather conditions of rain, flood, and bitter cold, your officers and men met a most searching combat test which required not only individual courage, skill, endurance, and determination, but also sound tactical judgement coupled with an insatiable desire to close with the enemy.

3. This achievement has added lustre to the glorious history of American arms, for which you and all the officers and enlisted personnel of your division and attached units, are highly commended.

G. S. PATTON, JR.,
Lieut. General, U. S. Army,
Commanding.

M-Sgt. Richard E. Clark

You were there.
Certified

M. McClure
Brig Gen U.S.A.

Patton gave the date for our attack. The 95th Division attacked Metz on November 8, 1944, a time when it had rained for two days straight and the Moselle River was at flood stage. Pontoon boats were used to cross the river at midnight, the artillery blasted the city, bombers also dropped bombs. The battle raged from November 8th to November 20th, for twelve days, when the German General surrendered, giving the 95th Division the nickname "The Mighty Men of Metz" and "The Bravest of the Brave," which later was made into a movie with James Whitmore by Hollywood. I, Sergeant Clark, received the Bronz Star Medal for my leadership in the Battle of Metz.

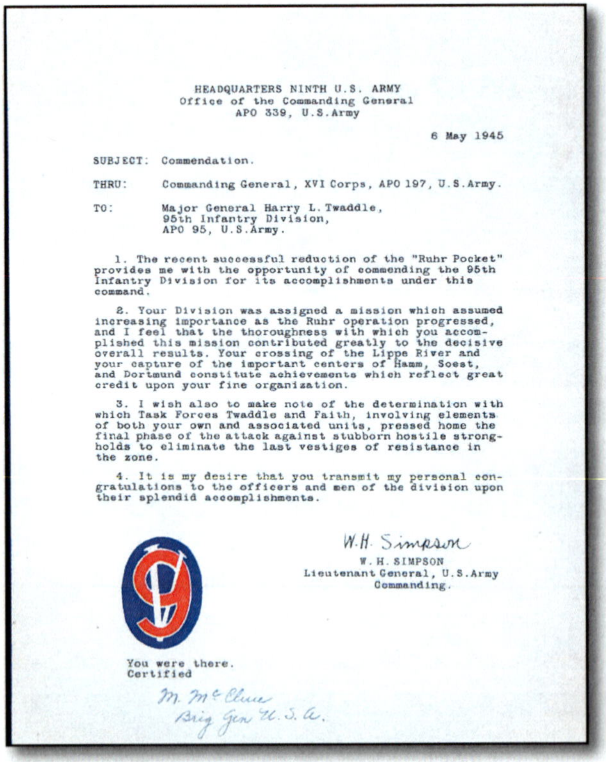

HEADQUARTERS NINTH U.S. ARMY
Office of the Commanding General
APO 339, U.S.Army

6 May 1945

SUBJECT: Commendation.

THRU: Commanding General, XVI Corps, APO 197, U.S.Army.

TO: Major General Harry L. Twaddle, 95th Infantry Division, APO 95, U.S.Army.

1. The recent successful reduction of the "Ruhr Pocket" provides me with the opportunity of commending the 95th Infantry Division for its accomplishments under this command.

2. Your Division was assigned a mission which assumed increasing importance as the Ruhr operation progressed, and I feel that the thoroughness with which you accomplished this mission contributed greatly to the decisive overall results. Your crossing of the Lippe River and your capture of the important centers of Hamm, Soest, and Dortmund constitute achievements which reflect great credit upon your fine organization.

3. I wish also to make note of the determination with which Task Forces Twaddle and Faith, involving elements of both your own and associated units, pressed home the final phase of the attack against stubborn hostile strongholds to eliminate the last vestiges of resistance in the zone.

4. It is my desire that you transmit my personal congratulations to the officers and men of the division upon their splendid accomplishments.

W.H. Simpson
W. H. SIMPSON
Lieutenant General, U.S.Army
Commanding.

You were there.
Certified

M. McClure
Brig Gen U.S.A.

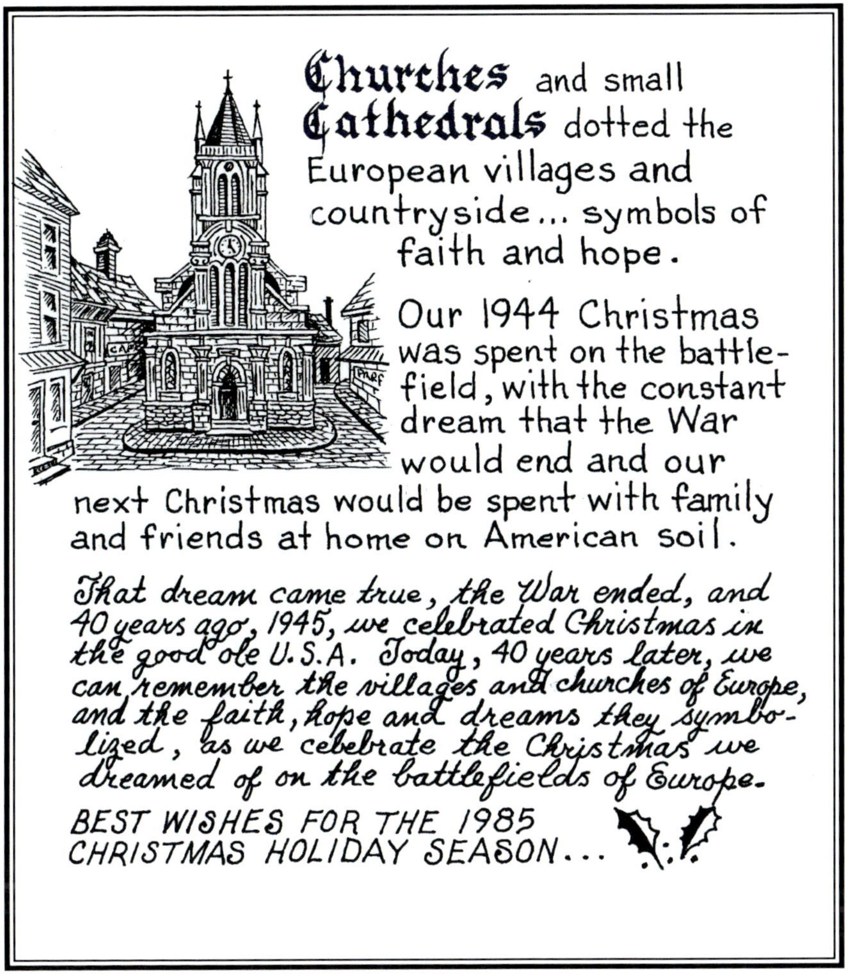

The 1985 Christmas Greeting drawn by Richard E. Clark
for the local 95th Division Veterans' Association

The drawing of the 95th Arty Metz Shield
by Richard E. Clark

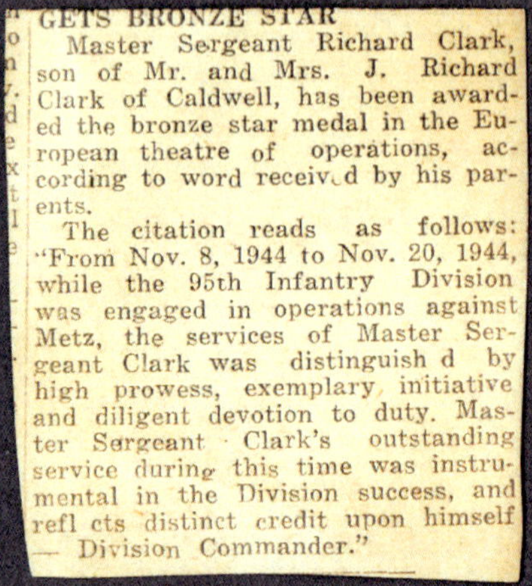

GETS BRONZE STAR

Master Sergeant Richard Clark, son of Mr. and Mrs. J. Richard Clark of Caldwell, has been awarded the bronze star medal in the European theatre of operations, according to word receiv.d by his parents.

The citation reads as follows: "From Nov. 8, 1944 to Nov. 20, 1944, while the 95th Infantry Division was engaged in operations against Metz, the services of Master Sergeant Clark was distinguish d by high prowess, exemplary initiative and diligent devotion to duty. Master Sergeant Clark's outstanding service during this time was instrumental in the Division success, and refl cts distinct credit upon himself — Division Commander."

After only eight days of rest, the Division was ordered to advance eastward and capture the city of Saarlautern, which was located on the Saar River and fortified by the Siegfried Line. Saarlautern was just as fierce a battle as Metz, involving hand-to-hand fighting. It took three weeks to take it. The 95th Division was victorious again, but received many losses in casualties, prisoners of war, and death."

Richard E. Clark, "Metz, France to the Saar River"

Chapter 7

Mopping Up

Entering Germany, France and Belgium

Winter, Felsburg, Germany

"We were stationed in this small village, which sat in a valley of surrounding hills. Note the German pillbox on the hill to the top left. Buildings were pretty much destroyed or damaged. In this area living quarters were on one side of the building, with an attached room or building where animals were kept (cow, pigs, and chickens). Note the manure pile in the front of the house and an inquisitive little boy watching an American soldier.

Some of us were living in this small shack near the Saar river. Note the trailer camouflaged with nets.

Richard E. Clark, "Winter, Felsburg, Germany Jan 8, 1945"

Richard E. Clark, "Winter, Germany"

Richard E. Clark, "Bombing - Kurl, Germany, World War II"

Richard E. Clark, "Bastogne and Liege, Belgium"

Following the capture of Saarlautern , the Division was sent north as backup troops at Bastogne and Liege, Belgium for the Battle of the Bulge.

This was a house with a roof blown off. At least we had a floor to sleep on instead of the ground. We had rounded up a pot-belly stove, which gave us some heat.

Richard E. Clark, "Living Quarters, Bastogne"

Richard E. Clark, "Funeral, Belgium"

Richard E. Clark, "Cemetery, Belgium"

Richard E. Clark, "Belgian Pine"

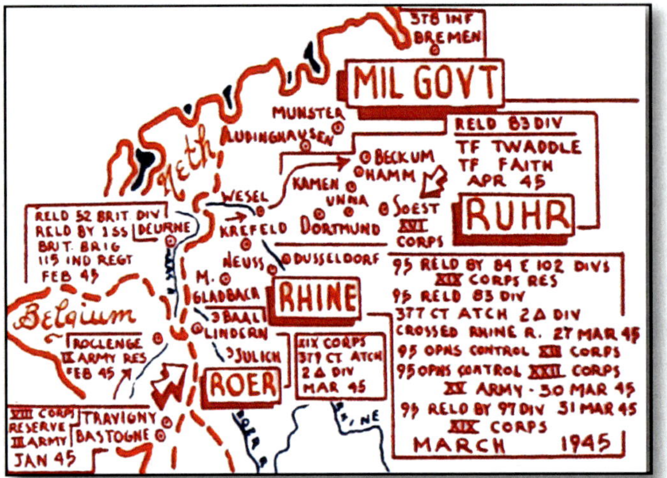

Richard E. Clark, "The Roer, Rhine, and Ruhr Rivers"

From there we crossed the Roer, Rhine, and Ruhr Rivers to the Ruhr area, mopping up in the industrial cities of Hamm, Soest and Dortman. A task force was sent North to the city of Bremen, the manufacturing city where German submarines (U-boats) where made. The Division was attached to the Great Britain 9th Army, at this time under the command of General Montgomery. I was in Liege where I saw old men wearing uniforms.

The war was drawing to a close by this time. The German army had withdrawn to almost Berlin. They were running out of gas. The British were forty miles from Berlin. They were taking everything from the West, East, North and South. The Russian army had advanced from the east and Berlin was surrounded. The war ceased, stopped. Citizens were all waving white flags and guns were held up over their heads. Hitler had committed suicide.

But the decision had to be made as to which country would take Berlin. The Allies or Russia (Stalin). All fighting was brought to a halt. A conference was held in Yalta attended by the three heads of the victorious nations, President Roosevelt, Winston Churchill and Josef Stalin. It all boiled down to who was the most forceful. There wasn't any doubt about who that was. Roosevelt was a sick man, drowsy and unsteady. Stalin was the aggressive one. Russia got the nod to take Berlin. The result of this decision had long-lasting effect; the beginning of the "cold war." Communism again raised its ugly head. Berlin was divided and the Berlin Wall was built, and Russia controlled many countries. We witnessed the results until President Reagan finally accomplished the removal of the wall dividing Berlin with the statement "Take down this wall." World War II in Europe ended at 9:00 AM, May 1st, 1945 (VE Day)."

Chapter 8

Homeward Bound

After the War

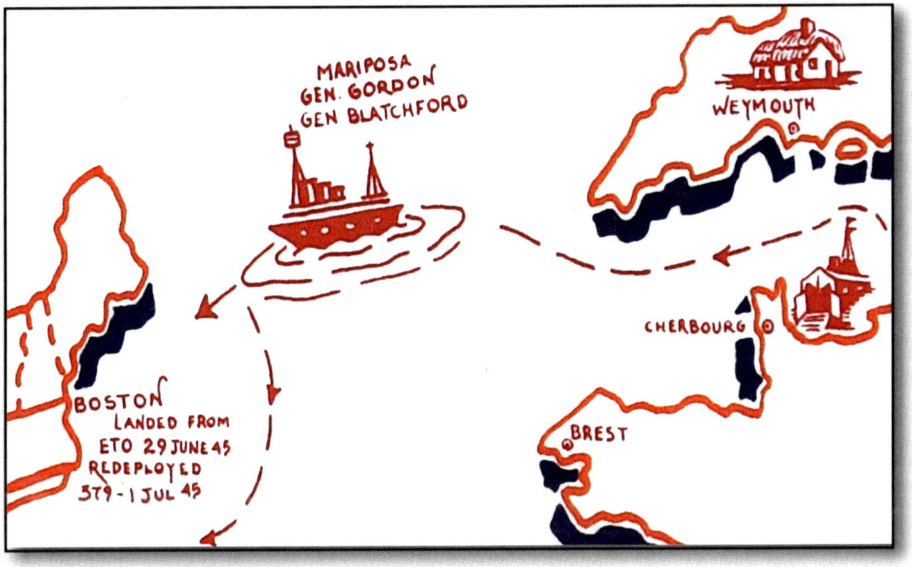

"From VE Day until the middle of June the Division was active in Germany serving as occupation and military police troops. June 29th 1945, the 95th Division sailed from Cherbourg France to the United States landing at Boston Harbor where the Division was redeployed to Camp Shelby Mississippi to prepare for embarkment to the Pacific area for combat with the Japanese. We were given thirty days furlough before we had to report to Camp Shelby. I went home to Caldwell, where my fiancé, Jessie Sherbourne, and I made plans to marry. She was working for the war cause in a Defense plant that made parts for PT Boats, in Canton, Ohio. We had a hectic day, getting blood tests, applying for a license, buying rings and driving back to Canton where we were married on July 3, 1945. (By the way that lady is here today, my wife, Jessie Clark.)

Richard E. Clark, "Miitary Identification Tags"

Mr. and Mrs. Richard Edward Clark
Wed on July 3, 1945 , Canton Ohio
(Jessie Marie Sherbourne)

From there we went as a married couple to Missouri to wait for my next deployment orders. When the news got out that the 95th Division was going to be shipped to the South Pacific, there was quite a rebellion. Wives, mothers, fathers and friends sent hundreds of letters in protest. A newscaster got wind of this news and conducted a campaign on the air in protest of the "Iron Men of Metz" having to serve more time in the war. They protested that they had done enough, suffered enough and were heroes. The Division was not sent. Two days before they would have departed however, the war with Japan ended when the United States dropped two atom bombs on Japan. Japan surrendered in September 1945 (VE Day). The 95th Division was deactivated October 15, 1945 and the veterans were sent home to be discharged."

Chapter 9

The Odyssey of the 95th Victory Division Map

Published by Richard Edward Clark

"In addition to drawings and paintings made during Mr. Clark's three years of military service, he kept a diary. This was not a written diary, but one rendered in art/map form. It was titled

"The Odyssey of the 95th Victory Division."

"The Odyssey" is a concise, summarized history of the Division's training, travels, locations of the division's maneuvers, supporting units, slogans of the Division ("Iron Men of Metz" and "Bravest of the Brave"), and sketches of the above. It is suitable for framing to hang in an office or den.

Mr. Clark was not permitted to publish the Art/Map while a member of the armed services. However, due to the hundreds of requests for a copy, after the war was over he returned to his home town of Caldwell, Ohio and he and his wife opened an art studio and devoted a year to publishing the history. It was copyrighted, published and sold in 1945 – 1946.

A copy hung in The War Department of the Pentagon Building and two enlarged copies currently hang in the National Memorial Museum in Oklahoma City, Oklahoma. As of the date of this publication, the 95th National Memorial was relocated in 2011, along with the Division headquarters , to the new Armed Forces Reserve Center at Fort Sill, Oklahoma.

Richard E. Clark standing in front of The National Memorial Museum in Oklahoma City, Oklahoma.

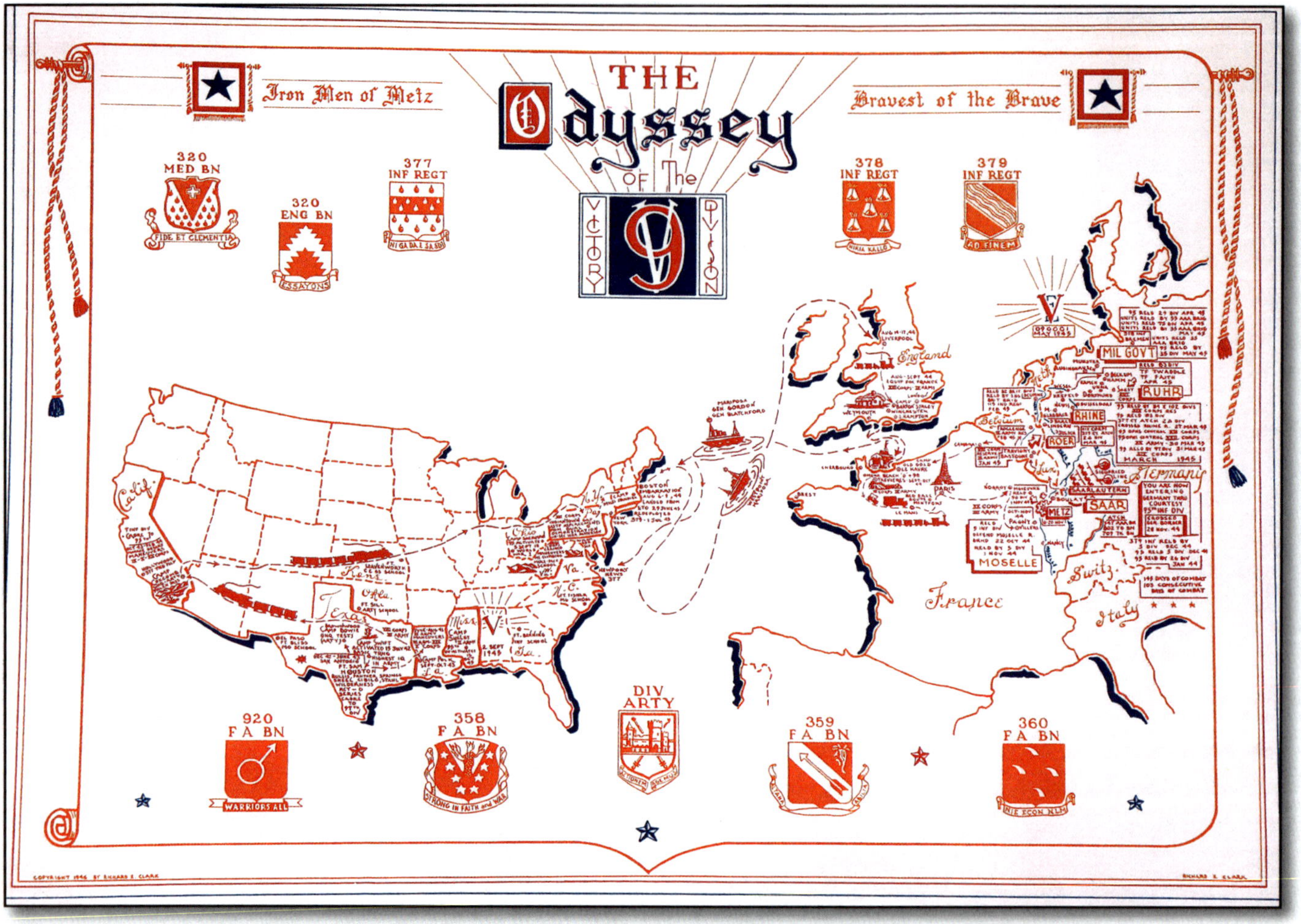

The Odyssey of the 95th Victory Division Map

Published by Richard Edward Clark

*Enlargements of Locations, Movement of Battalions, Battles, Dates and Illustrations from the Odyssey Map have been placed within the previous text to illustrate the areas as the information was given.

Personal Comments by Mr. Clark about

"The Odyssey of the 95th Victory Division"

"Although I had received requests for copies of the Art/Map History, there was no assurance that we could count on these as actual sales. We gambled, however, and spent most of our meager savings in the printing, promotion, publishing and mailing of the maps. We sent brochures to the known home addresses of the listed members of the Division. To our surprise, we received several thousand orders. We sent 25,000 hand addressed response letters to wives, parents, friends.

Although the results of our endeavor were gratifying, the greatest satisfaction came in many of the orders we received, which contained a letter from a mother whose son had been killed or a wife whose husband and been killed in action. We were the first people from the Division with whom they had any contact regarding their husband or their son. They asked us if we could please give them any information about their son's or husband's death – how killed, when, where, circumstances. Out of the thousands of soldiers in the Division, I didn't know their sons or husbands. However, we filed every order by battalion and company. In many cases we found the names and addresses of the deceased's buddies, which we sent to the mothers and wives (which required a great amount of time writing personal letters). The thank you letters received back from the mothers and wives saying they had contacted buddies of their son or husband and had received information they so much wanted, was indeed a great satisfaction for a service rendered... mission accomplished."

Richard E. Clark

Richard E. Clark at a lecture with a his "The Odyssey of the 95th Victory Division" Map

Through a Daughter's Eyes

As I was growing up, and in the years following, 'The Odyssey of the 95th Victory Division" always hung in its gold guild frame within our home. I recall that it hung in the midst of the framed awards and military medals, that I really never totally understood, and took for granted for most of my life. I cannot help but think now, of all the discussions I could have had, and the opportunities I could have pursued to learn so much more from my father. The Odyssey Map would have been the perfect catalyst for this. However, we do not realize some things until we ourselves are mature enough to look back and understand their significance. Today, the framed Odyssey of the 95th Victory Division map holds a very special place in my heart. And through it, I am getting to know my father and the reason that he shaped his life as he did. I realize the tenacity, sentimentality and patriotism that both he and my mother felt as they dedicated time from their lives in order to fulfill this dream for themselves and the families of the Mighty Men of Metz.

As a viewer, I find that by looking at the map, one is able to trace the travels, history, dates and significant events through pictorial images, that tend to make it more real, more relatable and more memorable. Viewing and consulting the Odyssey of the 95th Victory Division as I have put this book together, has been a great help. I hear my Dad's voice, with its Southeastern Ohio drawl, the clearing of his throat and the emphasis in his expression as he told the story of the map and its significance to not only himself and my mother, but to all of the soldiers and their families who lived much the same journey that he did during World War II.

Recently, I conducted an online search for The Odyssey of the 95th Victory Division Map and found a number of references to the map. Some were grandchildren who had come across it framed in their grandparent's homes; or others who found it still neatly rolled within the cardboard mailing tube. Some questioned "Is this something of value?" I assume their intent was monetary value. But I wonder, if it somehow led them to discoveries of their own, to come to know more about the sacrifices their loved ones had endured for our country. Perhaps they discovered a different sort of value about their loved one. For I know that those of the 95th Victory Division who lived to tell it, had much to tell about the experiences they lived during World War II. Without these stories, history cannot be truly understood and teach us the valuable lessons that should be so dear to our hearts.

Close-up Sections of
"The Odyssey of the 95th Victory Division Map"

Published by Richard Edward Clark

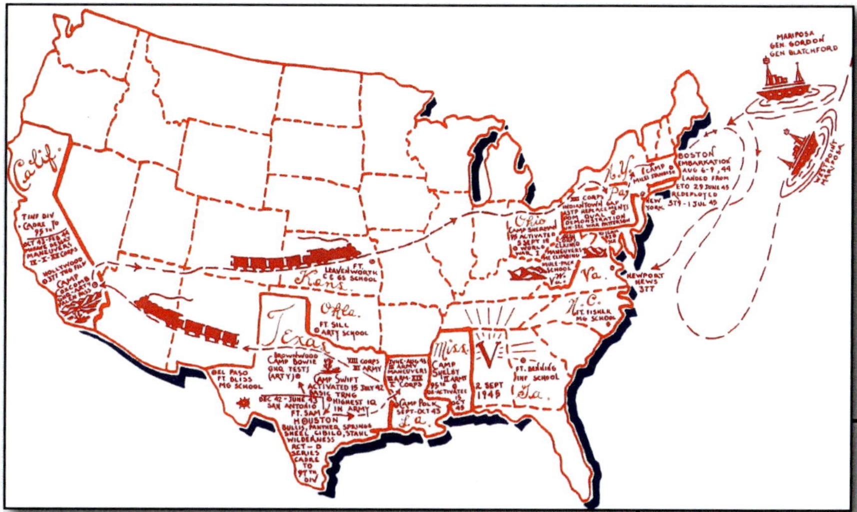

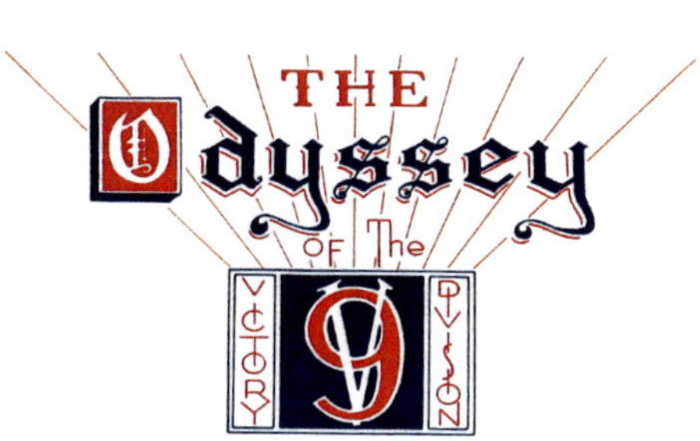

378
INF REGT

379
INF REGT

320
MED BN

320
ENG BN

377
INF REGT

The above image is the original oil cloth drawing of "The Odyssey of the 95th Victory Division Map." Drawn and designed by then Master Sergeant Richard Edward Clark.

Chapter 10

Return Renew

Through a Daughter's Eyes

After my father had spoken about the experiences of the war, he paused in his speech and said that he had two other things that he wanted to talk about. The first had been the story behind The Odyssey of the 95th Victory Division Map. The second had to do with his family. He then introduced those family members who were able to attend that very special event at Lima ArtSpace. He introduced his bride, help mate and wife, Jessie Clark. He also talked about his daughters whom he had raised in a home which was filled with art. He spoke with great pride as he told how both of his daughters, Becky (Clark)Latta and Lynda (Clark) Porter had been art teachers in the Lima City Schools. Becky had been an Elementary School Art Consultant and Lynda had been the Adaptive Art and Music Consultant. Both of us taught in the Lima City Schools while he was the Administrator of Art and Public Relations for the Lima City Schools. Jessie had also returned to school to earn her credentials and directed and taught Cosmetology in the Vocational Department at Lima Senior High School. Dad introduced Becky who was in attendance with her husband, Jerry, who was also an Elementary School Principal in Defiance, Ohio, (and a previous teacher at Horace Mann, and Assistant Elementary Principal at Garfield in Lima). He thanked Jerry for all the photographs he had taken of Dad's art work, and the video he was taking on this special day. He then introduced his granddaughters, Angie and Amy and her husband Dan who were in attendance that day. He mentioned his other daughter Lynda, and her husband Dan, and their two sons, Nathaniel and Benjamin. However, we could not attend, as we were living in North Dakota at that time and I had come down with the flu. I now have many regrets that I had not been able to attend this once in a life time event.

Dad continued with the second thing that he wanted to talk about:

"Five Years ago our daughter, Lynda Porter was living in Belgium. Lynda took art classes from me and taught art in Lima. Her husband Dan Porter worked for BP and became Board President of the Arts Council of Lima. They became Co-Chairs of Square Fair. Some of you may know them. They invited Jessie and I to come and visit them in Belgium. I wanted to see Metz, France. So we drove there to see it, and some other sights on the way. It was simply beautiful! We drove downtown one night for dinner. We crossed over the Mosel River. People were driving little cars everywhere. Dan pointed out the church which sat on an island which extended out into the river. Jessie took a picture of it, and did a painting of it later. I wanted to see the Monument to the 95th Division. The waiter drew a map for us and said "Merci Beaucoup, Merci Beaucoup! Thank you! Thank you very much! The monument with an American Eagle is for liberating our people." After dinner we went to see the monument and we took photographs and toured the tower and the fort area.

Cathedral of Metz

After spending time in Metz, we drove to Luxembourg City, Luxembourg. 10,000 soldiers are buried there. Each grave is marked by a single white cross, laid out in a semi-circle toward the apex, another white cross for General Patton.

Monument to the 95th Division
The Iron Men of the Metz
Pictured left to right: Nathaniel Porter, Jessie Clark, Benjamin Porter, Lynda Clark Porter, and Richard E. Clark

Richard E. Clark standing in front
of the Fortress of Metz

The Monument of the 95th Division
Pictured left to right: Daniel Porter and Jessie Clark

Chapter 11

"And Now for the Rest of the Story"

My father concluded his speech at ArtSpace with these words:

"As Paul Harvey would say "and now for the rest of the story." And there is a 'rest of the story". I'm not going to tell you the rest of the story. It's said that one picture is worth 10,000 words. The rest of my story is told in pictures. On the wall we've placed about twenty copy photos taken by an army photographer assigned to our Division. I've been pretty reserved in my description of my Division's journey and battles. These photos aren't so reserved. In fact, they may be pretty unpleasant to look at, even gross. But they tell the true picture of what the Division went through, and encountered along the way. They speak for themselves and tell the rest of the story.

As we observe Veteran's Day this year, I can't help but think of those 10,000 white crosses we saw at the American Cemetery in Luxembourg. 10,000 bodies and souls that didn't come home. 10,000 men, sons and husbands who paid the ultimate sacrifice, the last full measure of devotion fighting for their country and the free world.

There's an old saying that says "Be it ever so humble, there's no place like home." I'm sure there were thousands of veterans who while they were lying in their bedroll in their pup tents dreamed of their home and the time they'd be going home to the States. I know I did. I'd dream of my humble home in Caldwell. And it was a humble home. I grew up in the Depression. My Dad was out of work for two years. It was a very humble home. But it was my home; no other place like it.

All wars are terrible. World War II was terrible. But if it had to be – and I'm convinced it had to be; we were fighting against a man who may have bordered on being a genius. After all, Germany was soundly defeated in World War I. It was devastated, down trodden, the people were hungry, they needed a leader and the time was ripe for a young fireball named Hitler, who with his oratory and promises was striving to build Germany into a powerful nation of a supreme Arian race. The people were followers, brainwashed. But he made good his promise. He built one of the greatest war machines that ever existed. A machine that took a country about half the size of Texas and conquered all of western Europe; that furnished the manpower to fight on many fronts – in Africa, in Europe on their West, and Russia on the East. But he also bordered on being a mad man, a lunatic. He didn't have any qualms about torturing, murdering, gassing, burning the bodies of innocent people who weren't of his pure, supreme Arian race. This was massive ethnic cleansing. But he miscalculated. He didn't know the difference between right and evil. He didn't realize that the power of the spirit of democratic freedom was more powerful than his dictatorial war machine and all of his propaganda. He and the German people who followed him believed him and believed in his evil philosophy didn't learn from the lesson taught them by the defeat of Kaiser in World War I. He finally learned the hard way by committing suicide. What a disaster in the history of what could have been and is today a great nation- Germany.

Lynda Clark Porter
In Rememberance of 9/11/01

World War was a terrible thing. But I think it had to be. I think this war was justified. And if this war had to be, thank God it wasn't fought on our own shores and the streets of Lima. Had it been, probably many of us wouldn't be here today in our home sweet home of Lima, Ohio.

I'd like to make a pretty frank and controversial comment. I believe in freedom of expression. If anyone has seen the monument in Washington D.C. of soldiers raising our flag on Iwo Jima, or the picture of fire fighters raising our flag on a pile of debris in front of what was the World Trade Center in New York City, I simply can't understand how or why anyone could or would or might burn our flag in order to express themselves about their anger, frustrations, or desire to make a statement about our government and way of life, and the right to do so under the protection of the constitution for freedom of speech.

I have a request: If you happen to be where there is a Veterans Day Parade and our flag is passing by, or even if you are at a ballgame when the Star Spangled Banner is played or sung and the Flag is being hoisted up to the top of the Flag pole, if you're a Veteran, please salute. If you're not a Veteran, place your hand on your heart……. Long may it wave.

It's been a pleasure to be with you. I want to thank ArtSpace Lima for inviting me to show some of my art work and tell my story of the Odyssey of the 95th Division. You've been a very patient audience. Thank you."

Richard E. Clark
November 4, 2001

Richard and Jessie Clark standing in front of his map, "The Odyssey of the 95th Division"

Jerry and Becky Latta, Angie Thieszen, Richard and Jessie Clark, Amy and Dan Pflug at the ArtSpace in Lima, Ohio

Through a Daughter's Eyes

After my father concluded his talk, there was a moment of pause between himself and the audience. However, not long after there was a burst of applause as people sprang to their feet. As one can well imagine, there were strong relationships and ties my folks had established with many in this audience over a course of 38 years. My mother, sister and brother in law, their daughters and families were strongly touched and proud of Dad's words, as he peeled away layers of experiences that they had not fully realized. Pride and love were present and strong. There were handshakes, hugs, tears and pats on the back. I recall from watching the video that Dad, with an ear to ear grin on his face, introduced the girl he married, Jessie and his daughter, Becky and her family. Her husband, Jerry had once again been behind the video camera, recording the presentation as an historical and monumental event in our family's lives.

Although I was not there in person, I had been there in heart and soul. How I would have loved to share this experience with Dad as I had the trip to Metz. But it was a wonderful experience for my older sister Becky to share personally with her father and mother, and her own family. I could later observe from the video that there was much emotion within those in attendance, and particularly for my father. He had finally realized something come full circle, the revealing of the portion of his life that many did not know. It explained so much about who he was, and why he chose the paths that he did throughout his life. These were similar paths that some of those in the audience would have witnessed. A couple of men had been stationed near to Dad's journeys. For these men the impact was a visual walk back into their own memories of the formidable World War II experience.

After the presentation the guests were invited to view the exhibit of Dad's World War II Art Work, and The Odyssey of the 95th Division Map, as well as the accompanying photos from The 95th Victory Division Journals of World War II, of actual scenes, locations, battle grounds, destruction and reality. Dad had never wanted to highlight these, but in his speech, he reiterated the importance for future generations and those present to understand the reality of war, and fighting for freedom. For myself, as I viewed the video numerous times during the writing of this book, I found it, although aged and a bit garbled with unfiltered background noise of the other events happening at ArtSpace that day, to be a monumental occurrence for my mother and father. Perhaps it was a long over-due release of sorts for him as he finally told and relived his own version of 'the rest of the story."

Chapter 12

Additional Artwork During the War

In addition to the painted scenes and sketches Dad was able to create during the war, he was able to carve out some time for other artistic endeavors. These included illustrated newsletters during his basic training within various locations and training camps in the United States. These illustrations, appear years later in his artistic expressions in other medias and formats.

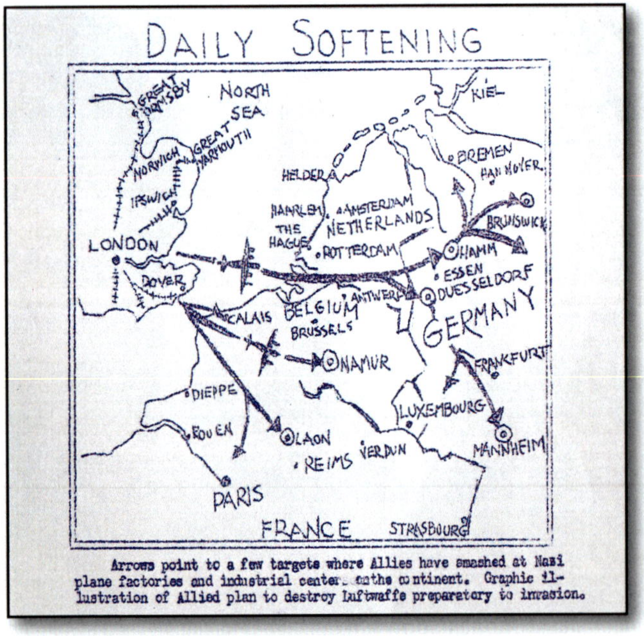

Arrows point to a few targets where Allies have smashed at Nazi plane factories and industrial center. onthe continent. Graphic illustration of Allied plan to destroy Luftwaffe preparatory to invasion.

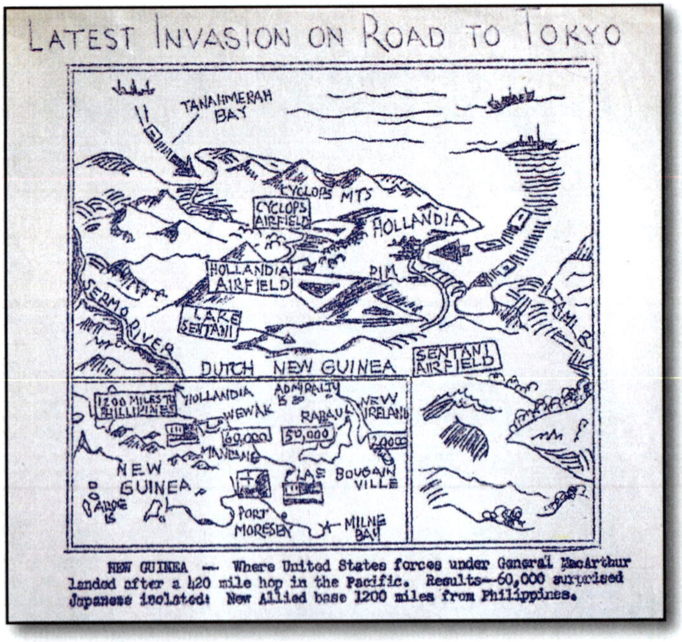

NEW GUINEA — Where United States forces under General MacArthur landed after a 420 mile hop in the Pacific. Results—60,000 surprised Japanese isolated! New Allied base 1200 miles from Philippines.

He also enjoyed painting serious portraits and tongue-in-cheek caricatures of soldiers and officers. These appeared to be cherished by those to whom they were presented. It is assumed that he was active in creating canvas maps that were disseminated to troops as well, but I am not certain of this assumption.

Richard E. Clark, "A Portrait of Brigadier General Ward Maris"

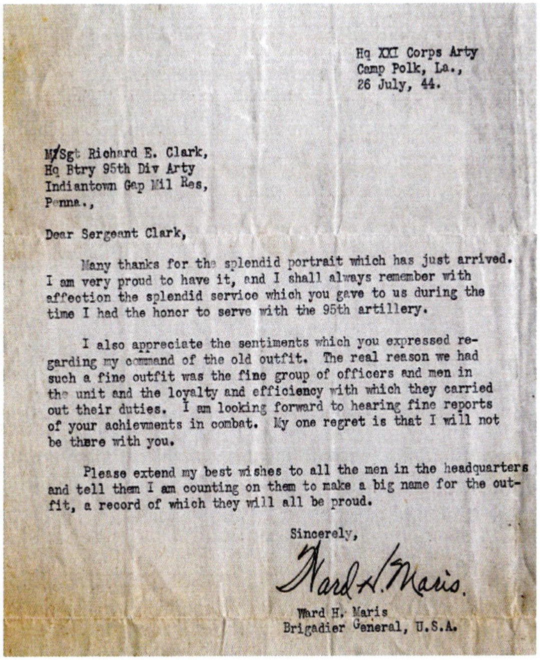

Hq XXI Corps Arty
Camp Polk, La.,
26 July, 44.

M/Sgt Richard E. Clark,
Hq Btry 95th Div Arty
Indiantown Gap Mil Res,
Penna.,

Dear Sergeant Clark,

Many thanks for the splendid portrait which has just arrived. I am very proud to have it, and I shall always remember with affection the splendid service which you gave to us during the time I had the honor to serve with the 95th artillery.

I also appreciate the sentiments which you expressed regarding my command of the old outfit. The real reason we had such a fine outfit was the fine group of officers and men in the unit and the loyalty and efficiency with which they carried out their duties. I am looking forward to hearing fine reports of your achievments in combat. My one regret is that I will not be there with you.

Please extend my best wishes to all the men in the headquarters and tell them I am counting on them to make a big name for the outfit, a record of which they will all be proud.

Sincerely,

Ward H. Maris

Ward H. Maris
Brigadier General, U.S.A.

Richard E. Clark, "The Odyssey of the 95th Division Map" inside the book of the 95th Division.

Chapter 13

After the War

The Legacy Continues

After my mother and father completed the year devoted to the Odyssey Map of Victory 95th Division, they were finally able to begin to put the pieces of their lives back together and move into the future with much determination and fervor. My Dad looked toward their future and the career goals that he had always had. He accepted a position as an art teacher in Celina, Ohio, a small town on the opposite side of the state. He taught there for a couple of years, and then made another move to Troy, Ohio - a bit farther southeast. While there, he began to develop and hone his teaching style, take on extra-curricular responsibilities and become involved in the community. Dad began taking his Art 4 students on field trips to the Cincinnati Art Museum, with stops at the Cincinnati Zoo and Botanical Gardens.

He formed Prom decoration committees, which involved the creation of large chalk murals of the selected Prom theme. In those days, nothing was thought of having groups of students come to the home. I have memories of groups of teenagers working on large pastel murals on large white paper taped to the walls of our basement. Popcorn, made and served by my mother and father, was an essential ingredient. My sister and I were able to be part of these fun events. Dad also painted large Christmas murals for the bank and town center traffic circle. He carried that further by including student participation. Students painted large Christmas scenes on the windows of the down-town shops and stores. By this time, Mom and Dad had built a new home, and Dad decorated the house in the same Holiday spirit. These later became a tradition that he and his daughters painted in our own home. We would place longnecked bendable wire desk lamps behind the windows to illuminate them for passers-by.

In his post war years Dad depicted social commentary in some of his art work, as he explored abstract art.

Dad branched out more into developing his public speaking abilities, as he became the PA announcer at the high school football games. We felt very special to be able to climb up all the clanging, echoing metal stairs and bleachers to visit our Dad in the glass fronted boxes at the top of the stadium. Dad and Mom both became more involved in civic and service organizations, particularly Lions Club International. Dad continued to develop his organizational, speaking and leadership skills. Mom initiated a Lions Tamer wives' group, not to mention raising two daughters, my older sister Becky and myself. My Mom and Dad were an amazing force together.

Dad began to pursue advanced degrees, with graduate work at Miami University of Ohio (my alma mater, as well) and obtained a Master's in Administration and Supervision Degree from Wittenberg in Ohio. These added degrees led him to accept a position as Art Supervisor and Lima Senior High, Senior art teacher in Lima, Ohio. Lima was about an hour or so North on I-75.

THE SIGNIFICANCE
of
CREATIVITY
. . . . in various fields
of man's endeavor

RICHARD E. CLARK

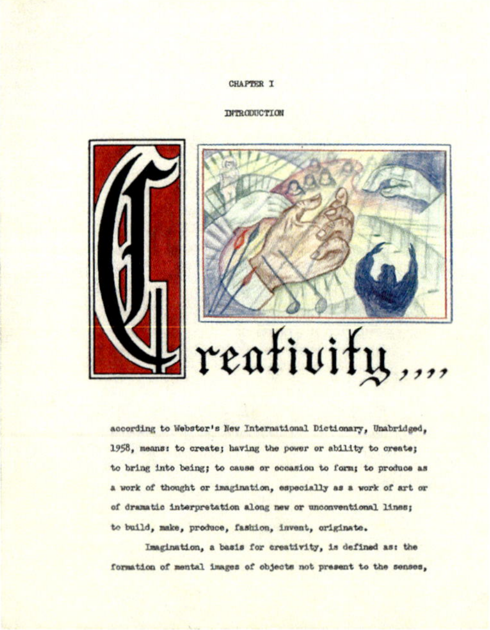

CHAPTER I

INTRODUCTION

Creativity,,,,

according to Webster's New International Dictionary, Unabridged, 1958, means: to create; having the power or ability to create; to bring into being; to cause or occasion to form; to produce as a work of thought or imagination, especially as a work of art or of dramatic interpretation along new or unconventional lines; to build, make, produce, fashion, invent, originate.

Imagination, a basis for creativity, is defined as: the formation of mental images of objects not present to the senses,

Richard E. Clark's Master's Thesis

As Becky and I moved into our high school years, we both were able to take art classes from our Dad during our Senior year. It was fun to say “Good morning Mr. Clark” when we passed in the hall or entered the art room.

Some of the images and seeds of concepts from his war time newsletter illustrations, reappear in the art work of this era, and in the Christmas letters done for the Ninety Fifth Division Chapter 7 Northwest Ohio Local Association.

During the Lima years, he continued his growth through teaching and advancing in his responsibilities to also include Director of Public Relations for the Lima City School System. He broadened the outreach of school involvement in events throughout the community, and served on numerous community and Arts Boards, and organized and ran School Levy Campaigns. He also grew in his roles within the Lions International and the Ohio Lions organization, and honed his public speaking abilities. He delivered High School Commencement Addresses, Memorial, Veterans and Fourth of July speeches throughout the State of Ohio, and conducted his own Public Speaking workshops within Lions International. He even ran for a public office, in Troy, before we moved to Lima. At the time that we had moved from Troy to Lima, Dad received a call from someone in Troy, encouraging him to run for Sheriff. We all laughed and laughed at this because in reality Dad had the stature, looks and mannerisms of Barney Fife. This was a mental image that none of us would ever forget, nor fail to chuckle about…Dad in a Sheriff's hat (appearing to be at least two sizes too big!)

Harkening back to his younger days, Dad designed numerous emblems for organizations or local celebrations, to serve the public as he included his love of art. He won the competition for the design of the Allen County, Ohio Bicentennial Emblem. He never stopped encouraging his students to use their art as a means to become involved in the community through arranging student art work exhibits in hospitals, malls, civic events and down town store displays.

FRONT OF MEDALLION…

BACK OF MEDALLION (Flexible color selection)

Dad never gave up his love of painting. He chose to give many of his paintings as gifts instead of selling them. He was an inventor of games and gadgets; a floating basketball hoop and an updated version of the old barrel ring hoop and stick game. In our home, art supplies were always abundant. His version of paper puzzle pieces that we colored with crayons; and then watched him assemble them into a piece of art, was a favorite pastime for my sister and I. Much later in life, I realized that this activity may have stemmed from the large mosaics of photos that he pieced together in his planning of maps, artillery routes and projections.Ironically, as I became an art teacher, I incorporated my own version of this creative puzzle into art activities for my special education students. My sister had created a similar activity in teaching art classes, and graciously shared the concept with me.

Becky and I enjoyed participating in my Dad's summer art classes back in his home town of Caldwell every year, including stops at the corner drug store for old fashioned cherry coca cola sodas. We were special guests on field trips to Museums and zoos with his students, Lions activities, parades and many vacations which were centered around traveling to the Lions Club International Convention in many locations across the United States. These ranged from LA, to New York City, Chicago and Atlantic City, New Jersey. I believe that our parents wanted us to see the beauty of our country, its freedom, its work ethic, and its patriotism. He knew the contrast first hand and wanted us to appreciate the gifts that freedom brings, and that our wonderful parents brought to us.

Richard E. Clark Teaching Summer Art Classes in Caldwell, Ohio

My parents, Jessie and Dick Clark, continued to touch lives through teaching and service to community, Lions Club, neighbors, The 95th Division Association, love of family and art. Upon retiring, when they were snowbirds in Tucson Arizona, Dad began to teach art lessons at the Recreational Center in their retirement village, as well as painting many large murals as backdrops for social events and theatre productions. The last time I visited their Park, Dad's painting of the community center still hung in a prominent place in the office. It was the first thing you saw as you came through the door. At some point, Mom decided to take a break from the many haircuts and styles she did for the park residents, and began to take lessons from the teacher she had watched throughout all the years, her husband. She began to create her own art works, and even branched out beyond the Clark traditional themes. Mom painted scenes for our sons to hang in their bedrooms, as well as Southwest and floral themes. One of her most beautiful paintings was of the Cathedral in Metz. It hung on their dining room wall as a constant reminder of that amazing trip. After Dad passed, Mom managed to take up the loose ends and teach Dad's art classes, that were so missed by the residents of their community.

The Art of Jessie Clark

The Art of Richard E. Clark

The Art of Richard E. Clark

The Art of Richard E. Clark

My parents' legacy continued, as Becky and I both pursued degrees in art education. We both were able to teach in the Lima City Schools. Dad became Becky's supervisor as she was an Elementary Art Consultant. I taught Adapted Art and Music, and although my job fell within the Special Education Department, my office was just around the corner from my Father's, until he retired. His surprise visit to one of my classes, was a meaningful highlight of my career. After starting families, and moving away from Lima, both my sister and I elected to stop teaching to raise our families. Becky pursued her love of wheel-thrown pottery and sewing. Her pottery pieces graced the shelves of shops and arts and craft fairs within North West, Central Ohio and Michigan for years. I plunged into painting, like my father, and illustrating Children's Books, written by my sister in law, Rosanna Porter. The love of art expression has been a constant companion, as our family had numerous moves due to my husband's career, including Belgium and England.

As is often said, "what goes around, comes around." The love of the arts has been a strong foundation within our marriages and our families. We have all come to understand what a distance our parents had traveled from their meager beginnings in the hills of Southeastern Ohio; a farm in a hollow below a winding dirt road, and a small town, nestled into the hills, where walking the Square on a Saturday night was the highlight of the week.

The Art and Pottery of Becky Clark Latta

Chapter 14

The Second Trip of a Lifetime

In October 1997, our excitement mounted as the boys and I waited at the Brussels airport with an oversized, crisp paper and raffia wrapped bouquet of brilliant autumn flowers. With anticipation, we finally saw the familiar, tired and a bit overwhelmed faces of Gramma and Grampa Clark appear through the crowd of weary travelers as they came through the doors into the terminal. They were here! They were here! "Bonjour!" we called out. We were so excited and relieved that they had managed the journey as we gave them huge hugs and presented the bouquet to Gramma Clark. We were very happy that they had come to spend time with us in Belgium and we were eager to share "the journey of a lifetime" with my parents. Although as sometimes things can happen at unfortunate times, our youngest son Ben had injured his leg on the playground at school, and I needed to take him to the Doctor that afternoon. We dropped Nate and my parents at home, which worked well to help them recover from the long flights. The rest of the day for Ben and I became a series of visits to various locations, as was the typical process in Belgium, for an exam, ex-rays, and the fitting of a soft cast. We learned that Ben had suffered a hairline fracture of his leg from a fall on the slide at school. Although, my parents did find much relief in having the time to rest, my mother then also slipped on our marble steps and scraped her leg. We were off to a complicated, yet determined start.

The Grand Place

Once we had moved past these initial events, we officially began the tour of Belgium. We enjoyed visiting the Grand Place in the heart of Brussels with its famous sites. We visited chocolate and lace shops and the famous "Mannequin Pis" statue of a small boy tucked into a cobble stoned lane just off the Grand Place. We sampled Belgian frites with mayonnaise, and "gaffres" (waffles). It was late afternoon after Dan was done with work for the day, when we headed down town. The buildings shone with golden light, the cobblestones glistened and the crowds were animated. I can still see my mother's face full of joy and enthusiasm as she spontaneously

Lynda Clark Porter, "Town Hall by Starlight, Brussels, Belgium"

marched alongside a small typical Belgian parade of uniquely, bright costumed figures on stilts. Memories flooded back to me of my mother marching in other parades. My parents were enthralled with the culture of the Grand Place! As I look back now, I realize that the contrast had already begun to take hold in my father…the contrast between his last memories of Europe, Belgium in fact; the horrors of war, the visions he witnessed, the plans he made, the maps he charted and the realities that ensued.

We continued with some local tours, including our favorite small German village, Monchau, not a very far drive from Brussels. We enjoyed wiener schnitzel, small shops, German music, pretzels, gingerbread, licorice and gazing over a stone wall at the bubbling stream. We visited Castle Beersel with its squawking, threatening geese; The International School of Belgium, where the boys and I spent much time; Waterloo, le Butte de Lion

Jessie and Richard Clark in Monchau Germany

and small local shops and restaurants. We also visited local patisseries, flower shops and drove through the famous Le Foret de les Soignes as we headed back to our own little village 'Centre". There we strolled through the local market set up on the plaza, and wandered near the pond and the elegant floating swans.

Lynda Clark Porter, "The Swans" from the book Tofu Ling

Once my parents had acclimated a bit and regained their land legs, we began our scheduled tour of the Belgian countryside, parts of France and Luxembourg City. Unfortunately, we hit another stumbling block after we drove south out of Hoeilaart. We had decided to take two cars so that there would be space for everyone with luggage, and two aching legs in tow. I was following Dan in my car when I heard a large 'clunk' and felt the power of the car immediately decelerate. We phoned Dan on our large cell phone with the bad news. Fortunately we were all able to pull off at the close-by exit and gratefully found a service station waiting for us. We soon learned that the transmission had dropped in my car. Therefore, with no other alternative, we all piled into Dan's car and we took off again on our tour, leaving my car at the repair shop just over the border into France, to be picked up on our way back. We decided to push our concerns about the repair to the back of our minds during the trip. We did not want to spoil this particular journey for my parents and did our best to keep spirits high.

Lynda Clark Porter, "Hoeilaart Town Hall"

From here we visited beautiful Dinant on the Mosel river. We had lunch and rode the Funicular up to the top of the rocky bank for the view and to visit the fort remains. Grampa remarked that this felt familiar, and I noted the wistful look in his eyes as he looked across the steep chasm at the citadel. We proceeded to Trier, where the boys purchased their favorite jester-like hats. From there we made our way to Metz, France. This was the most poignant journey of my parents' visit with us. I was not certain how this experience would affect both my mother and my father. I knew there would be apprehension about revisiting memories of war-torn lands, cities, towns, villages and most importantly, lives, families, faces, loss and devastation. I realized that the images retained in my father's heart were deeply embedded, and for a lifetime. But as we entered the beautiful town of Metz we passed the magnificent Cathedral of Metz, set on an island in the river. My dad perked up and even spoke a bit of French with a twinkle in his eye. "Parlais vous Anglais?" he asked the waiter at the small café where we stopped for lunch in the heart of Metz.

We mentioned to our waiter that Dad had been a soldier in the 95th Division during World War II. He smiled broadly and replied "Ah oui… The Iron Men of Metz. They hold a special place in our hearts and prayers

Richard Clark at the Memorial to the "Iron Men of Metz"

for the saving of our town, our families, our heritage and our future." He went on to explain that The Iron Men of Metz were honored in the city of Metz. He directed us to the Monument in honor of The Iron Men of Metz, and to the tower and fort area of town. At that moment, I realized and regretted that I had not set up an appointment for my folks to meet the Mayor and to have had a more auspicious welcome. But the look on my Dad's face said it all. This was enough. This was meaningful, and difficult, and joyous and memorable... all at once.

Jessie and Richard Clark at the Iron Men of Metz Memorial

We proceeded to the fortress of Metz that the 95th Division had cleverly and efficiently defended. They had managed to wall off the city below ground and thus thwart any further intrusion. These underground tunnels were sophisticated training, communication, strategic, German Officer Headquarters. They held the power of the German Forces. The Iron Men of Metz, 95th Victory Division had stopped it cold in its tracks. This was a turning point in the war. We all stood in amazement and overwhelming emotion as we viewed the monument, The American Eagle, next to the tower that honored The Iron Men of Metz. Although, I could see the emotion in his eyes, the pain he must be feeling, the memories of the reality he had witnessed, I listened intently to my father's words. He explained some of what had occurred, and yet I sensed there was so much more he could not say. He knew that they were appreciated, revered and honored by the people of Metz, France and Europe. He saw that they were acknowledged in many people's hearts for their bravery, sacrifice and fierce defense of freedom. My Dad's eyes glistened with tears and I knew that it took all he had to stifle the flood of sobbing emotion that was overflowing within his being. His eyes said so much as we took photos of him and Mom, the boys, and Dan and I in front of the Memorial. I could see the memories and the heaviness of the moment as we took his picture with his grandsons. There was so much contrast between the photos and the futures of these young boys, and the memories which he held within his heart.

In this moment, I realized that this was one of those moments that remain within your lifetime of memories and are locked within your own heart. This day must have held a significance beyond words and thoughts to my father as he realized a sense of life come full circle, and a revered task completed. He could touch and see palpable results of life renewed and restored to beautiful surroundings and the business of everyday life within a peaceful French community. If I could freeze this moment in time, I would cherish it forever, as I still do. And as I later came to realize, the fact that my father never told us, was that he had earned his Bronze Star for his exemplary contribution with his plotting, mapping and careful planning of artillery fire and routes, in the battle and winning of the Battle of Metz. He truly was an Iron Man of Metz, of the 95th Victory Division. After leaving Metz, we stopped in Luxembourg City on our way North to Hoeilaart, Belgium. We had planned this although we knew it would also be an emotional journey for my Dad. We began with lunch at an outside café which overlooked the deep rocky gorge. As we sat eating and enjoying the view, Nathaniel noticed that Ben had red spots on his face. We all looked at him- stunned to see this appearance. Ben had also mentioned that he was not feeling well, but I had assumed some of that had to do with his hairline leg fracture. We decided to proceed with our Luxembourg stop before making the last hour and a half drive North to Hoeilaart.(After we returned home, we learned that Ben did indeed have chicken pox. This added to our list of maladies and concerns that Dad might develop Shingles from exposure to Ben's Chicken pox. However, in Belgium, Doctors still visit homes, and that he did, while also cautioning us to keep Ben and Grampa a bit separated. Fortunately for us all, Dad did not get shingles.)

We all piled into the car and headed to The American Battle Monuments Commission, Luxembourg, American Cemetery where General Patton lies in rest, surrounded by the graves and crosses in honor of American soldiers. "The graves area contains 5,076 American military dead who lost their lives in the service of their country. The Pylons on the outside of the Chapel terrace name 371 missing in action of the United States Army and Air Force. 101 headstones mark the Gravestones of "Unknowns" whose remains could not be positively identified. White marble shafts surmounted by a Star of David mark the graves of 118 of the Jewish faith, and white crosses mark the others. The headstones are set in nine plots of fine grass. Separating the plots are two malls radiating from the memorial and two transverse paths. Two flagpoles overlook the graves area in front of the memorial. Between the flagpoles is the grave of General George S. Patton, Jr." (The information is from a pamphlet at the Luxembourg American Cemetery.)

The American Battle Monuments Commission, Luxembourg

The impact of this visit was again, palpable as my father stood silently in honor of rows and rows of white crosses and fallen soldiers. All of these held together in the center and supported by the white cross and grave of their leader, General Patton. I sensed his thoughts and prayers, his sorrows and his blessings, as the world around him must have stood still in time. The contrast, once again hit me. Yes, life in Luxembourg had been restored, yet so much life was lost not so many years ago. I later did learn the impact of this moment in time on my father, as I read his words about the visit to the American Cemetery in his talks at ArtSpace, Lima Lions Club and The Optimist Club about his World War II experiences. And I again regretted that I had not heard them in person.

The Belgium trip was not yet over, and there were more pleasant memories yet to be made with my parents; a weekend in Paris with their daughter was the last adventure. But this moment was the defining moment of a journey worthwhile and a dedication to my father who had helped the cause, plotted and planned, measured and mapped, and painted in the light of bombings against the night sky. He had survived. He lived to hold all the experiences deep inside.

Although, as the years after the war ticked by, those memories, lessons learned and philosophies of life had taken root. I believe that he made a choice to pass these on to his family, his friends, his colleagues, his service clubs and his communities through his art work, his teachings, his speeches, addresses and talks. He passed these on as he embraced life everyday with purpose and intent to make a difference.

Master Sergeant Richard E. Clark

Epilogue

A Look Back

Through a Daughter's Eyes

Years later, in 2015 my husband and I along with another couple took a riverboat cruise along the Rhine River. The cruise took us from Basel, Switzerland to Amsterdam, Holland. Prior to the cruise we did a driving tour which led us through the wine regions of Champagne, Burgundy, Alsace as well as visits to Bruges and Brussels, Delft Holland, and Lucerne Switzerland.

As we took them to the locations so familiar to us in Brussels and Bruges, I could not help but revisit the memories of our times there with my parents. My parents had thoroughly enjoyed the horse drawn carriages, quaint village center and handmade lace of Bruges. Once we were on board the riverboat, I also thought of my father, as I knew he had been in some of these regions during the war.

Before we left on the trip, I had quickly reviewed the Ninety Fifth Infantry Division History Book that I had found among the books my father had saved from his military service. There I found a page marked with a rusted paper-clip. It was marking the Adolph Hitler Autobahn Bridge on the Rhine River. I quickly read through the section of text across from the photo and learned that this bridge had indeed been a significant transition for the Ninety Fifth Infantry Division. I quickly made a copy of the map in the book and stuffed it in with my other maps and travel information.

Lynda Clark Porter, "Rhine River Region"

The bridge was just outside of Duisburg, Germany near Uerdingen, precisely where our riverboat would be gliding on its last leg of the cruise up to Amsterdam. I mentioned this to the Cruise Director, and he agreed that he thought this was the bridge that had more recently been replaced, yet again. In researching the bridge more, I learned that it had recently been renamed.

During World War II, after the Battle of Metz, 95th Division Battalions and Companies had progressed to this location near the Adolph Hitler Bridge on March 3rd, 1945. The Battalions pushed forward with periods of backing off due to resistance, with the goal of reaching the Autobahn Highway to the bridge. Three main Battalions and Companies progressed, squeezing and securing areas near to the bridge. Two plans had been established, one if the bridge had been seized and was passable; the other if it had been destroyed by the Germans. During the midnight hour, a patrol was sent to observe the situation at the opposite end of the bridge. The patrol discovered that it had been destroyed by the Germans and was impassable. In addition, the asphalt and bridge were set afire. The second plan was enacted. General Twaddle ordered the attack from several different Battalions and the mission was successfully completed.

On the afternoon of March 5th the last German resistance in the zone had ceased. However, the battles continued on into the darkness and it was reported that German forces were seen in boats trying to escape on the river. A search light was placed by the river and was successful in stopping the escape attempts. After this, there was more resistance and it was more difficult than the Battle of Metz, because they were now past the Rhine River and into Germany, not France. By April 2nd the whole Division had successfully moved to the East Bank of the Rhine, beginning a new phase of combat.

Sources sited that Germany now faced total defeat. German troops were showing signs of low morale and were dividing, breaking up and being severely defeated. This may very well have been the turning point of the war.

After the Battle of Metz, my father's Division then moved into fierce fighting at Saaurlauten. But after that they were utilized more in rebuilding activities. I felt that this must have aligned with the timing of the Activities of the 95th Infantry conquests along the Rhine River and the Adolph Hitler Bridge. The bridge was rebuilt in 1950 and named the "Krefeld-Uerdingen Brucke".

I mentioned my father's history to our travel companions and we determined that we would be passing under the new bridge, south of Duisburg around 11:00 PM that evening. They suggested that we should all go to the upper deck to glide under the bridge just before that time. As we settled into our upper deck seats, positioned for a view of the bridge, I noticed Frank walk up to the wheelhouse and speak with the Captain. As he emerged, the lights came brightly to life everywhere on the upper deck, setting us aglow in the darkness. A large brilliant spotlight was directed forward, into the waters ahead of the boat. The reflecting lights bounced and twinkled on the waves caused by the boat's slow movement forward. I now realized that Frank had spoken with the Captain and crew who had decided to give this occasion the emphasis it deserved, just for me. As we glided silently forward in the river, the distant lights of the bridge became more distinct. They seemed to grow from

Krefeld-Uerdingen Brucke

twinkling spots of light in the distance to more distinct orbs of intense white light as we sailed closer and closer. As we glided under the new, spectacular contemporary bridge I felt the history and impact of the moment. The row of lights along the top of the bridge glowed in the darkness as we silently approached the bridge, glided under it, and then passed on into the darkness of the river… inconspicuously and without ceremony to observers. I glanced back over my shoulder for one last look at the reflected lights, and the reflective history of the significance of this bridge.

Although my father had not been at the Adolph Hitler Bridge on March 3-5th 1946, at that moment I felt the closeness of him. I felt the sacrifice and the wisdom beyond his years that occurred because of all that he witnessed and experienced during his Service in World War II. The war years changed him, as it had changed everyone. I understood more about the fight that was well fought for freedom. And in those moments I also felt the artistic insightful eyes that I had inherited from my Dad, and how he was able to extract beauty from war torn lives, cities, small towns and very personal, unspeakable experiences. He saw his talent as a gift that told a story. It was a gift to share and to record history. He used that gift throughout his life to touch other's lives. Somehow, I knew he was there with me, gliding under that new shining bridge. His history had come full circle, I knew then that his story and his works of art would be shared with much love and honor in the hope that it might live on and touch others as it had so impactfully touched me.

Richard E. Clark standing with his 95th Division Award

IMPORTANT DATES - 95TH INFANTARY DIVISION

1. SAILED FROM BOSTON ON U.S.S. WESTPOINT - AUG. 6, 1944: LANDED AT LIVERPOOL, ENGLAND AUG. 14, 1944.
2. LANDED AT OMAHA BEACH, FRANCE - D DAY + 98 DAYS.
3. BATTLE OF METZ - NOV. 8 - NOV. 20, 1944 (12 DAYS).
4. CROSSED GERMAN BORDER - NOV. 28, 1944.
5. BATTLE OF SAARLAUTERN - DEC. 5 - DEC. 26, 1944 (21 DAYS).
   --- 145 DAYS OF COMBAT, 105 CONSECUTIVE DAYS OF COMBAT.
6. NORTHERN AREA IN BELGIUM AND GERMANY - FEB. THROUGH MAY, 1945.
7. WAR ENDED IN EUROPE - 9 A.M. MAY 1, 1945 - VE DAY, MAY 1, 1945.
8. SAILED FROM CHERBOURG, FRANCE FOR THE UNITED STATES -- JUNE 20 LANDED IN BOSTON - JUNE 29, 1945.
9. (TRAIN TO CANTON, OHIO - JULY 2: MARRIED JESSIE JUNE 3rd, 1945).
10. CAMP SHELBY, HATTIESBURG, MS - AUG. 1 - OCT. 15.
11. WAR ENDED WITH JAPAN - SEPT. 2, 1945 --- VJ DAY.
12. 95TH DIVISION DEACTIVATED - OCT. 15, 1945.
    VETERANS SENT HOME TO BE DISCHARGED.

The author of this book, Lynda Clark Porter has illustrated the following children's books:

Tofu Ling

Tofu Ling and the Carp Banner

Tofu Ling and the Angel

Tofu Ling Sets Sail

Hawaiian Peek a Boo

Note:
Tofu Ling and Tofu Ling and the Carp Banner
were nominated for the Caldecott.

Tofu Ling and the Carp Banner
won the Children's Moonbeam Book Award.

Book covers are on page 114.

These books are available at
www.RaisykinderPublishing.com

Tofu Ling
Written by Rosanna I. Porter
Illustrated by Lynda C. Porter

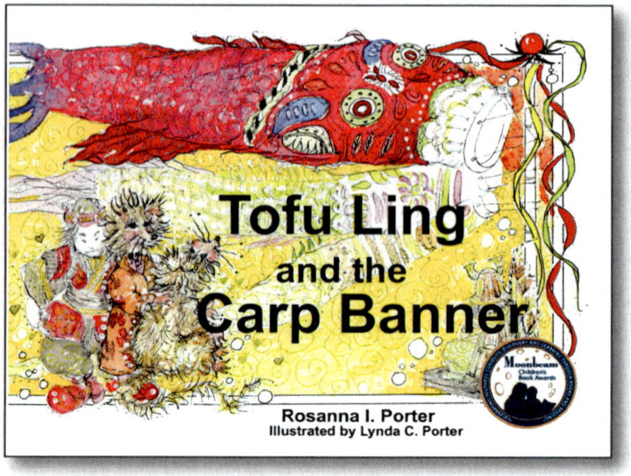
Tofu Ling
and the
Carp Banner
Rosanna I. Porter
Illustrated by Lynda C. Porter

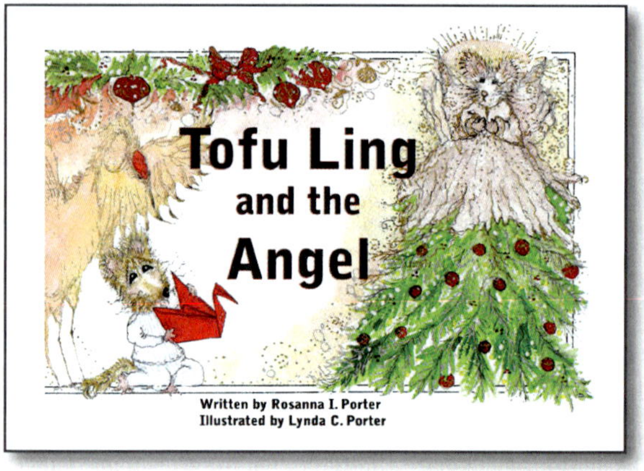
Tofu Ling
and the
Angel
Written by Rosanna I. Porter
Illustrated by Lynda C. Porter

Tofu Ling Sets Sail
Written by Rosanna I. Porter
Illustrated by Lynda C. Porter

Hawaiian Peek a Boo
Illustrated by Lynda C. Porter
Written by Rosanna I. Porter

Made in United States
North Haven, CT
04 November 2026

21726616R00064